Tim:

Thank you for your
leadership and all
of the support you
have given to my
efforts.

All the best.

Don

DOWNSCALING TECHNIQUES FOR
HIGH-RESOLUTION CLIMATE PROJECTIONS

Downscaling is a widely used technique for translating information from large-scale climate models to the spatial and temporal scales needed to assess local and regional climate impacts, vulnerability, risk, and resilience. This book is a comprehensive guide to the downscaling techniques used for climate data. A general introduction of the science of climate modeling is followed by a discussion of techniques, models, and methodologies used for producing downscaled projections, and the advantages, disadvantages, and uncertainties of each. The book provides detailed information on dynamic and statistical downscaling techniques in nontechnical language, as well as recommendations for selecting suitable downscaled datasets for different applications. The use of downscaled climate data in national and international assessments is also discussed using global examples. This is a practical guide for graduate students and researchers working on climate impacts and adaptation, as well as for policy makers and practitioners interested in climate risk and resilience.

RAO KOTAMARTHI is a chief scientist of the Environmental Science Division and department head for the Atmospheric Science and Climate research group at the Argonne National Laboratory. He applies numerical models to the assessment of climate-change impacts and uses high-performance computing and physics-based models for projecting changes at regional and local scales. His other research interests include the role of absorbing aerosols on radiative forcing and developing models for resource characterization of wind energy.

KATHARINE HAYHOE is a professor in the Public Administration program at Texas Tech University, where she is also director of the Climate Center. Her research focuses on developing and applying high-resolution climate projections to evaluate the future impacts of climate change on human society and the natural environment. She has served as lead author on key reports for the US Global Change Research Program and the National Academy of Sciences, including the Second, Third, and Fourth US National Climate Assessments and has been named the UN Champion of the Environment.

LINDA O. MEARNS is director of the Regional Climate Uncertainty Program and head of the Regional Integrated Sciences Collective at the National Centre for Atmospheric Research (NCAR). She has authored chapters in many of the Intergovernmental Panel on Climate

Change (IPCC) Assessment Reports, including the 2007 report that was awarded the Nobel Peace Prize. She is a fellow of the American Meteorological Society.

DONALD WUEBBLES is the Harry E. Preble Professor of Atmospheric Sciences at the University of Illinois. He is an expert in numerical modeling of atmospheric physics and chemistry, and has received the AMS Cleveland Abbe Award, the US EPA Stratospheric Ozone Protection Award, and the AGU Bert Bolin Global Environmental Change Award. He is a fellow of the American Association for the Advancement of Science, the American Geophysical Union, and the American Meteorological Society.

JENNIFER JACOBS is a professor in the Department of Civil and Environmental Engineering at the University of New Hampshire. She has more than five years of experience using novel weather and climate information to enhance infrastructure design. She directs the National Science Foundation–funded Infrastructure and Climate Networks (ICNet and ICNet Global) and was the lead author for the Transportation Sector Chapter of the Fourth US National Climate Assessment.

JENNIFER JURADO is Broward County's Chief Resilience Officer and director of the Environmental Planning and Community Resilience Division. In 2013 she was recognized by the White House as a Champion of Change for her work on climate resilience. She serves on the Board of Directors for the American Society of Adaptation Professionals and the American Geophysical Union's Thriving Earth Exchange.

DOWNSCALING TECHNIQUES FOR HIGH-RESOLUTION CLIMATE PROJECTIONS

From Global Change to Local Impacts

RAO KOTAMARTHI

Argonne National Laboratory

KATHARINE HAYHOE

Texas Tech University

LINDA O. MEARNS

National Center for Atmospheric Research

DONALD WUEBBLES

University of Illinois, Urbana-Champaign

JENNIFER JACOBS

University of New Hampshire

JENNIFER JURADO

Broward County Resiliency Office

CAMBRIDGE
UNIVERSITY PRESS

CAMBRIDGE
UNIVERSITY PRESS

University Printing House, Cambridge CB2 8BS, United Kingdom

One Liberty Plaza, 20th Floor, New York, NY 10006, USA

477 Williamstown Road, Port Melbourne, VIC 3207, Australia

314–321, 3rd Floor, Plot 3, Splendor Forum, Jasola District Centre, New Delhi – 110025, India

79 Anson Road, #06–04/06, Singapore 079906

Cambridge University Press is part of the University of Cambridge.

It furthers the University's mission by disseminating knowledge in the pursuit of education, learning, and research at the highest international levels of excellence.

www.cambridge.org
Information on this title: www.cambridge.org/9781108473750
DOI: 10.1017/9781108601269

First published 2021

Printed in the United Kingdom by TJ Books Limited, Padstow Cornwall

A catalogue record for this publication is available from the British Library.

Library of Congress Cataloging-in-Publication Data
Names: Wuebbles, Donald J., author.
Title: Downscaling techniques for high-resolution climate projections : from global change to local impacts / Don Wuebbles, University of Illinois, Urbana-Champaign, Rao Kotamarthi, Argonne National Laboratory, Illinois, Katharine Hayhoe, Texas Tech University, Jennifer Jacobs, Environmental Planning and Community Resilience Division, Broward County, Florida, Jennifer Jurado, Linda Mearns, National Center for Atmospheric Research, Boulder, Colorado.
Description: Cambridge, UK ; New York, NY : Cambridge University Press, 2021. | Includes bibliographical references and index.
Identifiers: LCCN 2020040201 | ISBN 9781108473750 (hardback)
Subjects: LCSH: Climatology–Data processing. | Climatology–Technique.
Classification: LCC QC874.3 .W84 2022 | DDC 551.60285–dc23
LC record available at https://lccn.loc.gov/2020040201

ISBN 978-1-108-47375-0 Hardback

Contents

Preface   *page* ix

1   Impacts, Adaptation, Vulnerability, and Decision-Making   1
     1.1   Assessing Climate-Change Impacts on Human Systems   1
            1.1.1   A Changing Climate and the Need for Assessment   1
            1.1.2   A Brief History of Major Impact Assessments   3
            1.1.3   Generating Climate Information for Impact Assessments   6
     1.2   Adaptation Strategies for Coping with a Changing Climate   7
            1.2.1   Adapting to Changes in Average Climate   9
            1.2.2   Adapting to Changes in Weather Extremes   11
            1.2.3   Adapting to Rising Seas   13
     1.3   Decision Making for Adaptation and Climate Information Needs   15

2   Global Climate Models   19
     2.1   The Need for Climate Models   19
     2.2   The Evolution of Climate Modeling   20
     2.3   Physical Processes in Global Climate Models   24
     2.4   Advances in Climate Modeling and Model Resolution   28
     2.5   Evaluating Climate Models Using Past Climate   31
     2.6   Archives of GCM Simulations   34

3   Assessing Climate-Change Impacts at the Regional Scale   40
     3.1   Climate Projections for Regional Assessments   40
     3.2   Climate Projections by Region   45
            3.2.1   Projected Changes for North America   45
            3.2.2   Projected Changes for Central and South America   47
            3.2.3   Projected Changes for Europe   50
            3.2.4   Projected Changes for East Asia   51

3.2.5   Projected Changes for South Asia   54
3.2.6   Projected Changes for Africa   56
3.2.7   Projected Changes for Australia   59
3.3   Regional Projections of Sea Level Change and Marine
     Temperature   61

4   Dynamical Downscaling   64
4.1   Regional vs. Global Climate Models   64
4.2   The Physics of Regional Climate Models   67
4.3   Outputs from Dynamical Downscaling Models   70
4.4   Workflow for Performing Dynamically Downscaled Simulations   73
4.5   Evaluation of Dynamical Downscaled Model Simulations   76
4.6   Availability and Use of Climate Projections from RCMs   79

5   Empirical-Statistical Downscaling   82
5.1   The Origin of Empirical-Statistical Bias Correction and
     Downscaling   82
5.2   Statistical Methods and Models for Bias Correction and Spatial
     Disaggregation in ESDMs   83
5.3   Statistical Methods and Models for Temporal Disaggregation
     in ESDMs   92
5.4   Evaluation of Output from ESDMs   96
5.5   Comparison between ESDMs   98
5.6   Availability and Use of Climate Projections from ESDMs   100

6   Added Value of Downscaling   102
6.1   The Concept of Added Value in Downscaling   102
6.2   Added Value from the Perspective of Scientists and
     Decision Makers   104
6.3   Added Value in the Context of Dynamical Downscaling   105
6.4   Added Value in the Context of ESDM   109
6.5   Comparing Statistical and Dynamical Downscaling   114
6.6   Research Needs to Further Determine Appropriate Use of
     Different Methods   117

7   Uncertainty in Future Projections, and Approaches for Representing
     Uncertainty   121
7.1   Identifying the Need for Quantitative Future Projections 121
7.2   Uncertainty due to Natural Variability   123

7.3   Scientific Uncertainty   125
       7.3.1   Climate Sensitivity   125
       7.3.2   Structural Uncertainty   127
       7.3.3   Parametric Uncertainty   128
       7.3.4   Accounting for Scientific Uncertainty   128
7.4   Uncertainty due to Human Choices   130
       7.4.1   Scenarios Used for GCM Simulations   130
       7.4.2   Addressing Scenario Uncertainty in Impact Assessments   132
7.5   The Relative Importance of Different Sources of Uncertainty   135
7.6   The Importance of Quantifying Uncertainty   136

8   Guidance and Recommendations for Use of (Downscaled) Climate
    Information   139
    8.1   Introduction   139
    8.2   Global Climate Model Selection   142
    8.3   Emission Scenarios   142
    8.4   Natural Variability   143
    8.5   Selecting Downscaling Approaches   143
    8.6   Use of the Different Downscaling Methods   145
           8.6.1   Comparing Statistical and Dynamical Approaches to
                   Generating High-Resolution Climate Projections   146
    8.7   Recommendations Based on Particular Variables, Questions
           Asked, and Physical Characteristics of the Region   152
           8.7.1   Useful vs. Usable   152
           8.7.2   Climate Services and Web Portals   153
    8.8   Conclusion   155

9   The Future of Regional Downscaling   157
    9.1   A Look at the Future   157
    9.2   Future Directions for Global Modeling   158
    9.3   Future Directions for Regional Modeling   161
    9.4   Future Directions for Empirical-Statistical Downscaling   162
    9.5   Will Downscaling Become Obsolete?   164
    9.6   Coupling with GIS and Other Tools   164

References   166
Index   188

Preface

This book grew out of a report we prepared in 2016 for a US funding agency that primarily supports work related to environmental issues of concern to the Defense Department. The motivation for the report was Dr. John Hall, who was the program manager for the Strategic Environmental Research and Development Program (SERDP). The report was developed to address concerns that many of the users we expected to lead the effort on climate impact and adaptation had regarding how to evaluate, select, and use various climate-downscaling techniques. We thank Dr. Hall for that and getting us together on this journey.

This book is thus intended to be a practical guide to the process of using downscaled climate projections for planning and decision-making throughout the world. Nonspecialists with technical backgrounds in planning, engineering, risk assessment, and management should be able to use this book to obtain an understanding of the tools and techniques used to develop, analyze, and use climate projections for a range of practical applications. Graduate students interested in careers related to climate adaptation and preparedness and faculty teaching classes on climate change and climate-change impacts on the natural and built environments will benefit from the material discussed in this book.

Finally, we also hope that government employees in the areas of environment (scientists and program managers), defense and national security (long-term strategic planners, installation-level environmental managers, infrastructure planning, and management teams), and infrastructure, health, and commerce will find this book a helpful companion in their journey through the climate-change vulnerability assessment and adaptation process.

While we have attempted to cover as much of the general topic area of downscaling climate models and their use, we are certain there are many more topics that were not included in the book. Some recent publications on statistical downscaling techniques and their applications, mesoscale meteorological modeling, and global-scale climate models that go into more detail on the physics and numerical methods used for developing the downscaling techniques discussed, will be great companions to this book and should be consulted by readers seeking deeper understanding of the models. International assessments, such as those conducted by the IPCC and national-scale assessments from various countries around the world, discussed in Chapter 3, are great additional resources for all the topics discussed here and we highly recommend consulting these for country-specific challenges from the changing climate.

Finally, we wish to thank Emma Kiddle and Sarah Lambert of Cambridge University Press for their patience as we struggled to move from concepts to words on a page and guiding us to the finish line. All of us are also thankful to our home institutions, which gave us the freedom to work on this project, especially near the end of the project as we focused our energy to complete this book and were mostly missing from action at work.

1

Impacts, Adaptation, Vulnerability, and Decision-Making

Climate change is a broad-reaching, global challenge. It impacts most human and natural systems, from agriculture and ecosystems to energy and health, and exacerbates other preexisting issues, from poverty to political instability. Evaluating these impacts and our vulnerability to them increases awareness of the need for adaptation and resilience. This chapter provides a brief history of impact assessments, focusing on the models, tools, and information that are needed and are available to quantify future impacts across a wide range of systems and scales, and to provide valuable input to adaptation and resilience planning.

1.1 Assessing Climate-Change Impacts on Human Systems

1.1.1 A Changing Climate and the Need for Assessment

Throughout much of human civilization, current and past climates have jointly served as a reasonably reliable predictor of conditions to be expected in future decades. What types of building to build; when to plant and harvest; how to plan for the chance of winter ice, summer rain, recurring flood or drought – all of these could be estimated based on accurate records of the past. The historic record of drought could reliably inform water-supply planning, and historical rainfall distributions could provide a solid basis for drainage-system requirements.

Since the beginning of the Industrial Revolution, however, increasing emissions of carbon dioxide, methane, and other heat-trapping greenhouse gases have been altering average climate conditions at local to global scales in ways that are not reflected by historical records. Atmospheric levels of carbon dioxide are now higher than they have been at any time in at least the last three million years, and continue to increase (Hayhoe et al. 2017). Global average surface temperatures have risen in response, by 1.0 °C from 1880 to 2016. Through the end of the century, global temperatures are projected to continue to increase by at least another 0.6–1.0 °C if rapid and decisive action is taken to reduce and eventually eliminate heat-trapping gas emissions, and up to 5° C or more if not (USGCRP 2017; 2018b).

Already, changes in many aspects of climate at the local to regional scale, including average temperature, precipitation, and the frequency and intensity of extreme weather events, are affecting human society and the natural environment (Stocker et al. 2013; USGCRP 2017). The impacts of climate change are being experienced in sectors ranging from energy supply and infrastructure to agriculture and ecosystems (Field et al. 2014; USGCRP 2018). Increasingly, as the sea level rises and water levels change, precipitation patterns and seasonality shift, temperatures rise, and extreme weather events become more severe and/or more frequent, human society is experiencing the need for transformational change in how we plan and design for the needs of communities, regions, systems, and beyond.

The projected direct impacts of climate change range from perceived nuisances to severe disruptions, affecting water and wastewater systems, energy supplies, ecosystems, agricultural operations, coastal resources, transportation, national security, and the built environment. For example, even if average annual rainfall remains constant, a shift toward more intense rainfall requires additional capacity in stormwater systems to protect against flooding. Basic levels of service standards are also being called into question, such as where it may no longer be technically or economically feasible to maintain dry roads or even dry ground-level stories in buildings under more intense rainfall, but rather to reconsider a community's tolerance for more frequent roadway or even substructure flooding. These issues are of growing interest, relevance, and concern to both the public and private sectors, given the diversity of services and sectors affected.

It is not only the direct impacts on individual sectors that are of concern. Compounding impacts can act across sectors, challenging systems such as densely populated cities where community livability, economics, public safety, and public health are all interrelated and are all affected by a changing climate. In addition, while many impacts from warming or changing precipitation patterns are local or regional, the effects of local climate disruptions can ripple through international supply chains to impact international manufacturing, economics, transportation infrastructure, and more. Given the extent and potential magnitude of these impacts, a collective shift toward holistic, cross-sectoral planning is needed to ensure resilient communities and economies. With each power failure, wildfire, algal bloom, or flood, there is a reminder of the connection between the environment and our communities, our reliance on critical infrastructure, and the cascading impacts of severe weather and other climate disruptions across our society and beyond.

1.1.2 A Brief History of Major Impact Assessments

Box 1.1
The IPCC

Scientific assessments are essential tools that link the state of scientific understanding with the decision-making process. Assessments survey and synthesize science within and between disciplines and across sectors and regions. They highlight key knowledge that can improve policy choices, and identify significant gaps in the science that can limit effective decision-making. Assessments also track progress by identifying changes and trends in issues of concern, advancements in science, and developments in the human response.

The Intergovernmental Panel on Climate Change (IPCC) was formed in 1988 by the World Meteorological Organization (WMO) and the United Nations Environment Program (UNEP). As an intergovernmental organization, membership of the IPCC is open to all member countries of the United Nations (UN) and WMO. Currently, 195 countries are members of the IPCC. The purpose of the IPCC is to bring together experts from around the world every five to seven years to synthesize the most recent developments in climate science, adaptation, vulnerability, and mitigation. Interim reports, such as the special report on the impacts of global warming of 1.5 °C, are also used to examine specific topics. Governments request these reports through the intergovernmental process; the content is deliberately policy-relevant, but steers clear of any policy-prescriptive statements. Government representatives work with experts to produce the "Summary for Policy Makers" (SPM), which highlights the most critical developments in language accessible to the world's political leaders. The reports also undergo extensive peer review throughout the process before publication.

IPCC AR5 is the most comprehensive synthesis to date. Experts from more than eighty countries contributed to the three volumes of this assessment, with more than 830 lead authors and review editors and over 1,000 contributors. About 2,000 expert reviewers provided over 140,000 review comments. The AR5 assessment was more extensive than previous assessments in evaluating the socioeconomic impacts of climate change and the challenges for sustainable development. The inclusive process by which IPCC assessments are developed, reviewed, and accepted by the member nations ensures an exceptional level of scientific credibility. For this reason, AR5 serves as the primary basis worldwide to inform domestic and international climate policies. Many countries draw upon the IPCC in their national climate assessments; the United States does so in its national climate assessments. The 6th IPCC Assessment Report process is now underway, with the next series of assessments due in 2022.

Understanding the role of heat-trapping gases in the atmosphere, their influence on global temperature, and the extent to which both are increasing as a result of human activities, has been a topic of scientific study since the 1800s (Weart 2015). It was not until 1979, however, that the first formal assessment of climate was published. Called "Climate Dioxide and Climate: A Scientific Assessment" and informally referred to as the Charney Report after its lead author, eminent meteorologist Jules Charney, the report focused primarily on the science, ending with the simple but ominous statement that "it appears that the warming will eventually occur, and the associated regional climatic changes so important to the assessment of socioeconomic consequences may well be significant, but unfortunately the latter cannot yet be adequately projected." (Charney et al. 1979)

A decade later, in 1990, the newly formed IPCC produced its first Assessment Report (AR). Since then, the IPCC-ARs have included reports from three working groups. Working Group 1 focuses on the science, documenting past and future changes and the causes of those changes (see for example Stocker et al. 2013); Working Group 2, on observed and projected impacts, vulnerability, and adaptation in human and natural systems at the regional scale (see for example Field et al. 2014); and Working Group 3, on mitigation pathways, policies, and technologies to reduce the human impact on climate (see for example Edenhofer et al. 2014).

Since that first IPCC report, an increasing number of assessments have been published that document observed climate changes and summarize future projections for various cities, regions, and countries around the world (e.g., Australia: BOM and CSIRO 2012; 2014; 2016; 2018; Canada: Warren and Lemmen 2014; Bush and Lemmen 2019; see Skelton *et al.* 2017 for a full review). Some assessments go further, synthesizing and quantifying the potential impacts of

---

Box 1.2
The US National Climate Assessments

The US Global Change Research Act (GCRA) was signed into law in 1990 by President George H. W. Bush. This law established the US Global Change Research Program (USGCRP), with the aim that USGCRP develop and coordinate "a comprehensive and integrated United States research program which will assist the Nation and the world to understand, assess, predict, and respond to human-induced and natural processes of global change."

The GCRA also requires a report be prepared and submitted to the President and Congress every four years that (1) integrates, evaluates, and interprets the findings of USGCRP; (2) analyzes the effects of global change on the natural environment,

Box 1.2 (cont.)

agriculture, energy production and use, land and water resources, transportation, human health and welfare, human social systems, and biodiversity; and (3) analyzes current trends in global change, both human-induced and natural, and projects major trends for the subsequent 25 to 100 years. This report is referred to as the US National Climate Assessment (NCA).

Under the direction of USGCRP, which comprises thirteen federal agencies, the NCA helps individuals, communities, cities, regions, and the Federal Government prepare for the challenges of climate change. The First National Climate Assessment, entitled *Climate Change Impacts on the United States: The Potential Consequences of Climate Variability and Change*, was published in 2000. This assessment raised awareness and began a national process of research, analysis, and dialogue about the coming changes in climate, their impacts, and what Americans can do to adapt to an uncertain and continuously changing climate.

The Second National Climate Assessment, entitled *Global Climate Change Impacts in the United States,* was published in 2009. A much more comprehensive analysis was done for the Third National Climate Assessment (NCA3), which was published in 2013. The National Climate Assessment and Development Advisory Committee (NCADAC) was a sixty-person US Federal Advisory Committee that oversaw the development of the draft NCA3 report and made recommendations about the ongoing assessment process. The NCA3 report was written by more than 300 authors from academia; local, state, tribal, and Federal governments; and the private and nonprofit sectors, selected based on expertise, experience, and ensuring a variety of perspectives.

While the NCA is legally required, the process of producing and reviewing the NCA has varied between the different assessments. For example, NCA2 was written by a small team of experts, and was based in part on twenty-one different specialty reports on different climate-related topics. NCA3 relied on a large external committee that advised the author team, while the NCA4 was led by a Federal Steering Committee with representation from the same thirteen agencies that have a seat at USGCRP. In addition, each of these assessments has undergone an extensive external-review process, including public, National Academy of Sciences, and government-agency reviews.

With the NCA3, USGCRP also established a sustained assessment process that has resulted in a series of specialty assessments led by different US agencies. As a result, regional and sectoral activities are ongoing and special reports are produced on a more frequent basis. The reports currently published include: Global Climate Change, Food Security, and the US Food System (Brown et al. 2015); The Impacts of Climate Change on Human Health in the United States: A Scientific Assessment (USGCRP 2016); Effects of Drought on Forests and Rangelands in the United States (Vose et al. 2018); and The Second State of the Carbon Cycle Report (SOCCR-2) (2018).

climate change on relevant natural and human systems (e.g., the European Union: Cammalleri et al. 2017; the United Kingdom: Jenkins et al. 2009; UKCIP 2018; the United States: Mellilo, Richmond and Yohe 2014; 2USGCR P 2017, 2018). At the national scale, for example, the United States produces a congressionally mandated NCA at regular intervals that includes discussion of the state of the climate and a review of impacts by region and sector (see Box 1.2).

1.1.3 Generating Climate Information for Impact Assessments

As the extent and diversity of acute and chronic impacts on the natural environment and human systems expand, so too does the need for access to trusted tools and information to help inform adaptation plans, vulnerability assessments, and resilience planning. For many sectors and applications, reliable information regarding the direction of future change and, in some cases, the rate at which change is occurring and its expected future magnitude, is essential to developing robust, multi-decadal planning and management strategies.

Although different studies may have different national, regional, and sectoral focuses and may use different methods to assess impacts, most begin with the same premise: that future climate conditions under a set of consistent assumptions regarding human choices, as reflected in future greenhouse gas emissions, atmospheric concentrations, or radiative forcing, can be simulated by global climate models (GCMs; see Chapter 2). However, as first noted by Gates (1985), there is a fundamental mismatch of spatial, and sometimes even temporal, scale between the output of GCMs that tend to simulate climate at relatively coarse spatial scales and the information typically needed for assessing impacts on and determining adaptation measures for most human and natural systems. Thus, most applications require climate projections that are not only generated by GCMs but that are also *bias-corrected* and *downscaled* in some manner (see Chapters 4 and 5).

Downscaling refers to methods for developing regional or local information from coarser resolution information, usually generated from GCMs. In the earliest climate-impacts research, future climate projections from GCM outputs were bias-corrected and downscaled using the simple "delta method," whereby observations for a given location are scaled up or down based on the average changes projected by the GCM for that larger area (Mearns et al. 2001). Since IPCC-AR2 (1995), different approaches to downscaling, including the use of both higher-resolution dynamical modeling and statistical methods (see Chapter 8), have emerged. Today, a broad range of dynamical and statistical downscaling methods are available to increase the spatial, and sometimes even the temporal, resolution of GCM output.

Dynamical downscaling uses a high-resolution climate model centered on a relatively small region (ranging from a small region to a continent) that is driven by GCM output fields at its boundaries. Most high-resolution regional models use pre-calculated GCM output fields to update their boundary conditions every three or six hours, depending on the temporal resolution available from the GCM output. An additional dynamical approach is variable-resolution modeling, which involves using a global model that has coarser resolution around the world, but higher resolution over a particular area of the globe. Dynamical downscaling is discussed further in Chapter 4.

Empirical statistical downscaling establishes a statistical relationship between GCM output for a past "training period," and observed climate variables of interest that are then used to bias-correct and downscale both historical and future GCM simulations to the same scale as the initial observations, which can be either point sources or gridded observations. Statistical bias correction and downscaling are discussed further in Chapter 5.

A wide range of downscaling approaches is used to assess climate impacts, depending on everything from the logistical constraints of the analysis to the characteristics of the system being studied. Nevertheless, the majority of information on future projections and impacts originates from GCMs that have been bias-corrected and downscaled in some manner, reflecting the growing need for higher-resolution information on projected climate change (see Chapter 3).

While downscaling methods have been available for many decades (Giorgi and Mearns 1991) and have been used for producing future climate information for use in many impacts and adaptation studies (IPCC 1996; Parry and Carter 1996), there has been a lack of clear research to demonstrate the best practices for using these various methods. Unpacking the different methods, describing their origins, laying out their strengths and weaknesses, and providing guidance for their use in adaptation, vulnerability, and resilience planning is the focus of this book.

1.2 Adaptation Strategies for Coping with a Changing Climate

How can communities, organizations, regions, businesses, and countries plan for adaptation in a changing climate? In practice, many planners are frustrated by the limits of the climate information available for their region and by the extent to which most climate models are unable to represent the variables that matter most to them at the desired spatial and/or temporal scales. It is not only a matter of scale: often, climate-model outputs must be translated into the variables or indicators already used as input for planning – return period, threshold exceedances, degree-days, streamflow, or more. This requires that a nontrivial amount of time and effort be invested in communication and collaboration between experts in climate information and experts in quantifying the impacts on a given system.

For some, the issue of uncertainty in the climate projections is a challenge. The difference between certain observations (e.g., it is certain that the drought of record happened) vs. uncertain projections (e.g., the drought of record may become longer, but models suggest a large range of uncertainty in that change) poses yet one more troublesome impediment to adaptation planning. In reality, however, the uncertainty in future planning arises from multiple factors (see Chapter 7) and, depending on the system, climate-related uncertainty may not be the dominant, or even one of the most important, factors.

In the absence of absolutes, practitioners emphasize scenario-based planning, an approach that encompasses a range of conditions and timelines for evaluating vulnerabilities and adaptation options, accounting for uncertainty in multiple aspects of the system, from technology to human response to climate projections. Still, scenario-based planning does not supplant the need for improved downscaling and translation of GCM outputs to a level that is "accessible" and relevant at the scale of regional and local planning, while also working to further resolve the processes influencing some of the key parameters for improved scenario-based planning.

Box 1.3
Extreme Heat

In the port city of Da Nang, Vietnam, safety thresholds for heat indices are already routinely exceeded. The output from five different climate models that participated in the CMIP5 were downscaled using an empirical quantile-based bias correction and downscaling technique (see Chapter 5) for the lower RCP4.5 and higher RCP8.5 scenarios. These high-resolution projections were used to understand how heat impacts to Da Nang's workers and small businesses will increase in the future, and to develop strategies to reduce health risks. Like many coastal cities, Da Nang's topography changes from sea level to mountains over a short distance, resulting in localized wind and temperature patterns within the city and temperatures in the city that are about 5° C higher in the city than in surrounding areas. Because this variability in wind velocity and temperature all occurs within the spatial scale of a single GCM grid cell, statistical downscaling was used to better understand extreme heat changes in Da Nang, A key prediction is that the existing hot season will lengthen by two to three months between 2020 and 2049 and that the cooler nighttime temperatures that currently allow the population to recover from daytime heat will dramatically decrease over the same interval (Opitz-Stapleton *et al.* 2016).

Communities are already grappling with the challenges of translating climate modeling science into information that can be used to alter planning standards, infrastructure design, management plans, and more. These data and their interpretation will inform standards for major public-works projects, private investment, natural-system-management plans, and infrastructure intended to serve a community over the evolving conditions of the next half century or more. The value of these resources and their associated infrastructure is measured in trillions of dollars; so, although climate science is evolving rapidly in response to this need, nonetheless many affected communities cannot afford further delay. Regulatory standards are outdated and land-use decisions are in play. In an era of rapidly changing environmental conditions, informed and effective decision making will necessarily rely upon the ability of planners to access and integrate evolving, geographically relevant, climate science in regional and local planning: whether using the direction of the trend to inform no-regrets strategies, or directly linking downscaled climate-model output to sector-specific models to quantify local impacts of climate change and to develop strategies to address challenges to changing system stressors. As the 4th US National Climate Assessment documents, this adaptation is already occurring: just not fast enough to keep pace with the current rate of change.

1.2.1 Adapting to Changes in Average Climate

Global mean annual and seasonal temperatures are already warming; this warming trend is projected to continue. Precipitation is also changing, with some regions getting drier while other regions are getting wetter. In northern regions, warming conditions are causing the snowpack to decrease, a function of a later snow season and earlier seasonal melt.

A warming climate will shift agricultural growing seasons and agricultural zones, which will influence crop selection and yields, and affect livestock production. Warming is also causing a northward expansion of non-native species and insect-borne disease outbreaks. Changing climate patterns have been shown to shift terrestrial, marine, and freshwater systems. Terrestrial animal communities have shifted ranges an average of 3.8 miles per decade (Parmesan and Yohe 2003); shifts of up to 17.4 miles per decade have been recorded for marine communities (Cheung et al. 2009). Rising temperatures degrade air quality, particularly over already polluted areas. Rising temperatures, coupled with increased CO_2 concentrations, can influence plant-based allergens, hay fever, and asthma. Under future warming, wildfires could significantly increase due to an increase in the length of the fire season and an increase in warmer and drier days, with wide-reaching impacts on air and water quality (Liu et al. 2010).

As communities prepare for long-term shifts in the timing, amount, and type of precipitation a watershed may experience, water managers are employing advanced hydrologic models to assess impacts of climate change on water resources and their ability to meet water demand. Spatial and temporal changes in precipitation, rising temperatures, sea level rise, and competing demands for traditional water sources are all factors in these models. And supply and demand for energy and water are intimately linked; changes in climate that put pressure on water supply and demand can also drive an increase in energy demand. Specifically, additional energy is needed to deal with water-supply diversification, advanced water treatment, wastewater disposal, and increased flood management. Similarly, the higher temperatures that drive additional energy demand also have the potential to reduce the capacity and performance of energy-supply systems. For example, when the volume of available cooling water decreases or the inlet temperature of cooling water increases, the capacity of power plants that rely on a steady supply of cooling water may be reduced.

Adaptation to long-term changes in average conditions is already taking place across many sectors. For example, forest-resource management strategies are being modified to take account of climate change by including alterations to stand-density management practices, reducing surface fuel, controlling invasive species, and restoring aquatic habitat. In agriculture, there is a transition to faster-growing varieties of crops that can reach maturity more quickly, an adaptation to shorter and more intense growing seasons. Farmers in areas with predicted increases in water shortage are shifting to less water-intensive crops and heartier varieties, and are considering investments in alternative water sources for supplemental irrigation; on the other hand, wetter conditions are supporting transition to more water-intensive crops in historically semiarid regions (Huang et al. 2016).

Where warming conditions impact populations, adaptation actions include developing comprehensive response plans to extreme heat, climate-proofing health care infrastructure, and implementing integrated surveillance of climate-sensitive infectious diseases. Energy utilities are already proactively working to address the issue of projected increases in system load through improved demand management involving automation and smart controllers, including the addition of renewable and clean energy sources to traditional electrical sources. In the future, energy utilities will need to take active steps to improve demand management and water-supply diversification, similar to the efforts of water providers, and some are already doing so, with water-conservation improvements as part of infrastructure upgrades, use of reclaimed water for cooling operations, and investments in alternative water sources (e.g., South Florida).

Understanding and responding to local and regional impacts from a warming climate and changing precipitation patterns requires an understanding of changes that have already occurred and the magnitude and rate of change anticipated in the future. Thus, a thorough familiarity with observed trends, global-climate-model projections, and their limitations is requisite to address effectively the long-term changes via operations and management, planning, and the development of capital-improvement plans are needed to ensure sustainability and resilience under a changing climate.

1.2.2 Adapting to Changes in Weather Extremes

For many systems and regions, the earliest and even possibly more severe impacts of a changing climate are experienced through changes in the frequency, intensity, and/or magnitude of extreme events. These include more intense summer heatwaves, longer droughts, more frequent heavy downpours, and stronger coastal storms and hurricanes (Hayhoe et al. 2018).

Extreme heat can cause roads to melt, railways to buckle, public transit to overheat, and energy infrastructure to be taxed, leading to rolling brownouts or even blackouts. Natural systems, including stream and lake waters, can warm to the extent that they are no longer able to support important species such as salmon or brook trout and/or become prone to harmful algae blooms. Extreme heat also negatively impacts livestock and agricultural production. Humans experience heat stress and heatstroke when exposed to extreme temperatures; this disproportionately impacts children, the elderly, and the infirm. In addition to risks to health, outdoor workers may also experience a reduction in productivity. Urban

---

Box 1.4

Extreme Weather Risk Assessment to Transportation Sector

In 2013, Southeast Florida participated in a Federal Highway Administration (FHWA) assessment of transportation-infrastructure vulnerability to the effects of climate change and extreme weather. This assessment factored in the combined impacts of sea level rise, storm surge, and heavy rainfall. The study employed the FHWA framework and scoring system for rating vulnerability based on exposure, sensitivity, and adaptive capacity of railways and road networks. Regional datasets allowed comparable analyses among the various transportation links, integrating high-resolution topographic information and elevation data, rectified flood levels, and future-condition storm simulations. Low elevations, lack of redundancy, and long detours were common characteristics of the most "at-risk" segments. Recommendations emphasized formal planning for sea level rise in project design, redesign of drainage to handle more flow, and hardening of key infrastructure against extreme weather-related stressors.

populations are particularly sensitive because cities trap heat, creating heat islands that are considerably warmer than surrounding regions. Globally, many urban areas lack cooling mechanisms in residences or businesses, making their inhabitants particularly vulnerable to heat-induced health issues.

In some regions already at risk of flooding due to their topography or urban development, an increase in the occurrence of extreme downpours could increase flooding, prompting flood-management operations that preclude surface-water storage for water supply. For other regions, more frequent and severe droughts are expected to stress crops, deplete wetlands, harm riverine fisheries, and reduce groundwater available for drinking water. Changes in storm intensity, concentrated snow melt, and increased runoff present additional flood risk for communities. Above- and below-grade transportation systems are at increased risk from flooding and degradation, which reduces expected service life. Floods that cause transportation disruptions along the supply chain also limit food mobility. Increased precipitation extremes elevate the risk of surface runoff and soil erosion, resulting in degradation of freshwater and marine ecosystems due to increased sediment and nutrient loadings.

The combination of droughts, floods, storm surges, wildfires, and other extreme events stress nations and people through loss of life and displacement of populations. The interactions among these systems must be considered in addition to the effects of these stressors on individual systems. Failures can cascade from one system to another (e.g., where failures in physical infrastructure systems have downstream consequences for human health and safety). Extreme heat and floods can affect the diverse network of railways and roadways connecting communities, and the operations of seaports and airports, impeding local traffic and global commerce. Impacts on livelihoods can exacerbate conflict through intermediate processes, including resource competition, commodity price shocks, and food insecurity. Such disruptions have the potential for severe economic impacts, affecting not only the global flow of goods and services, but large-scale economic development, and a complex array of commercial, social, and community activities.

Increasing awareness of susceptibility to changing climate extremes has led many organizations, communities, and regions to implement adaptation measures, including using high-resolution climate projections to assess the viability of adaptation strategies. Solutions are diverse, influenced by local conditions and geographic context, and reflect place-based decision-making. For example, the observed increase in flood exposure and risk has already driven many communities to review and update flood maps and infrastructure-design standards, while investing in an array of improvements to protect communities from increased flooding. Some communities are updating building standards to require additional freeboard above-base flood elevations, elevating or reinforcing critical infra-structure, or updating stormwater plans to address increased flood hazard. Other examples include the preservation and use of green space for stormwater storage or even accommodation of flood waters through retreat under conditions of repeated inundation and increasing flood risk.

Energy utilities are also already proactively working to cope with increasing peak system loads by improving demand management through automation and the installation of smart controllers, and by adding renewable and clean energy sources to traditional electrical sources. In additional to utility-scale investments, institutional investment in combined heat and power projects, solar and storage, distributed renewable energy (e.g., roof-top solar), and micro-grids, provide a means of reducing peak demand and improving energy security on the individual, local, and regional scales.

In the health sector, early warning and response systems can protect population health from extreme heat today and provide a basis for more effective adaptation to future climate change. Adaptation efforts outside the health sector can have health benefits when, for example, infrastructure planning designed to reduce ambient temperatures and attenuate stormwater runoff. Projections of climate-change-related changes in the incidence of adverse health outcomes, associated treatment costs, and health disparities can promote understanding of the ethical and human-rights dimensions of climate change, including the disproportionate share of climate-related risk experienced by socially marginalized and poor populations.

1.2.3 Adapting to Rising Seas

Sea level is rising as a result of the thermal expansion of the ocean and the melting of land-based ice, primarily from the Greenland and Antarctic ice sheets (Sweet et al. 2017). The rate of sea level rise is accelerating, indicating that historical trends will systematically underestimate projected change. Total sea level rise by 2100 is projected to be as much as 1–2 meters (IPCC 2013; USGCRP 2017; 2018b) on average, modified by local trends in land uplift (as is occurring in many places in Alaska) or subsidence (as is occurring in the Gulf Coast). This sea level rise, occurring globally at a rate faster than any experienced in the history of human civilization, threatens coastal communities, infrastructure, and ecosystems.

It is estimated that around 10 percent of present-day population worldwide lives in the Low Elevation Coastal Zone, which is also home to two-thirds of the world's largest cities (McGranahan et al. 2007; Lichter et al. 2011). Countries with the greatest number of people at risk are primarily in Asia, including China, India, Bangladesh, Vietnam, Indonesia, and Japan; however, 40 percent of the US population also lives in coastal counties, making it number eight on the list of most at-risk countries from sea level rise (NOAA 2013). Coastal zones also house residential and commercial buildings and transportation infrastructure, from ports and navy bases to airports, roads, bridges, rail lines, and evacuation routes (Moser et al. 2014). They support a trillion-dollar tourism industry, and provide the livelihood for many subsistence fishers and farmers in low-income countries. There is no comprehensive global estimate of the value of infrastructure in the coastal zone yet; but by 2100, the costs of raising existing dikes and flood damage alone are estimated to range 1.5–2.5 percent of global GDP per year (Jevrejeva et al. 2018).

Sea level rise is already producing chronic high-tide and storm-surge flooding in many low-elevation regions including such places as the Southeast United States, Indonesia, and Bangladesh. Rising seas are spilling over seawalls, flowing through stormwater systems, and eroding coastal landscapes. As seawater occupies more space, reductions in soil storage and stormwater-system capacity increase the frequency and magnitude of storm-induced flooding under smaller-scale rainfall events. Coastal ecosystems, including mangrove forests, coral reefs, salt marshes, and seagrass beds, have lost more than 50 percent of their area due to anthropogenic and natural factors, including sea level rise. Their continuing degradation will reduce the ecosystem services they provide, including habitat for fish, recreation, flood protection, wave-induced erosion reduction, and carbon sequestration.

In some locations, planning, management, and building standards are beginning to be amended in accordance with projected sea level rise and hydrologic conditions. Adaptation strategies largely include the construction of barriers to keep seawater at bay, elevating and armoring of infrastructure, active flood management, and changes in land use, including retreat. Large-scale adaptation plans are also unfolding, with features that include comprehensive coastal storm barriers, expansive coastal wetlands, and open areas for surface-water storage.

Adaptation to sea level rise also occurs inland as regional flood, water-supply and transportation systems suffer cascading impacts. For many regions, adaptation will be managed incrementally, with regular and planned investments in response to system changes, such as the gradual loss of drinking-water wells to saltwater contamination, which can be offset through development of alternative water supplies over longer planning horizons. Planned retreat is also a potential adaptation strategy for communities that might identify a tipping point where further engineering in coastal adaptation may be ineffective or too costly. Some communities in Alaska and in the western Pacific islands are already planning such retreats. Coastal-zone management requires knowledge of current and future sea levels at a scale that is relevant to the infrastructure or ecosystem system at risk.

Communities and ecosystems will experience different levels of sea level rise and storm surge because mean sea level and storm vulnerability varies by location. Relative sea level (RSL) rise in this century will vary along coastlines due, in part, to a combination of sea level variations and vertical movement of the land. For example, it is likely that the RSL rise will be greater than the global average in the US Northeast and Southeast coastal areas and in the western Gulf of Mexico. Changes to the frequency and intensity of severe coastal storms such as tropical and extratropical cyclones (ETCs) also vary regionally, with impacts dependent upon the storms and their interactions with regional topography. While the inland propagation of rising seas and severe storms has a greater immediate disturbance in low-elevation areas, the loss of cliffs through erosion in steep coastlines may leave other regions defenseless. Vulnerability to sea level rise based on population and assets at risk are extremely high in parts of Asia, the eastern United States, and the Netherlands. Protection of coastal assets and ecosystems requires high-resolution information on RSL that also considers coastal storms and other climate-driven processes that could affect and alter the coastal environment.

1.3 Decision Making for Adaptation and Climate Information Needs

Creating and implementing adaptation plans requires making decisions under conditions of uncertainty. As discussed further in Chapter 7, climate-related decisions require the combined knowledge of possible future climate conditions, socioeconomic information, and relevant decision-making protocols. While most adaptation planning requires some information about future climate, the level of detail required can be varied (Dessai and Hume 2004; Dessai et al. 2017). Methods such as Decision Scaling (Brown 2011) and Robust Decision Making (Lempert et al. 2010) make use of future climate information in a way that allows uncertainty about the future to be accommodated in the decision-making process. These methods avoid the "predict-then-act" approach that tends to require a great deal of detailed climate information, highlighting how consideration of the decision-making protocols to be used in adaptation planning is important information for determining how detailed the future climate information needs to be.

In addition, adaptation decisions are made at different spatial scales. Local decision making (e.g., for the city of New York) may require more detailed future climate information than a more general plan for the state of California. Similarly, plans for city of London (Walsh et al. 2013) include detailed urban functions as opposed to the one developed for the UK (UKCP18 2018). For New York City's adaptation and resiliency planning (NYC Mayors' Office of Recovery and Resiliency 2019), future climate information was obtained from a large number of GCM simulations downscaled using the very simple "delta" method using two different scenarios for the future, the lower RCP 4.5 and the higher RCP 8.5 scenarios (Horton et al. 2015). In contrast, Ganguli and Coulibaly (2019) used output from three dynamically downscaled models that participated in the NA-CORDEX to develop intensity, duration and frequency curves for hydrological systems in Southern Ontario region. As computational resources become more easily available, the use of more complex methods for generating high-resolution climate projections, such as dynamic downscaling and advanced statistical techniques can be expected to become more prevalent.

Box 1.5
Coastal Resilience

In Southeast Florida, climate change is already impacting regional hydrology, drainage, and flood-protection systems. Low land elevations, flat topography, and dense coastal development present unique challenges to flood management and adaptation planning. A complex system of gravity-operated canals and flood gates provide drainage and stormwater management for the region's nearly 6 million residents. Designed in the 1940s, many of the control structures are operating at design capacity. Rising seas restrict stormwater discharges during seasonal high tides and the rise in the water table reduces soil storage capacity for rainfall infiltration, contributing to flooding.

In the Greater Fort Lauderdale area, Broward County government has formally adopted a scenario-based sea level rise projection (reflecting combined IPCC and federal agencies in the United States) as the basis for policy and planning. Areas of likely inundation with 2 feet of sea level rise are now delineated as Priority Planning Areas within the County's land-use plan, subject to additional review as part of any proposed land-use amendments. The County's 100-year flood map is being updated to reflect future flood elevations under the conditions of 2-feet sea level rise and intensification of predicted rainfall through the years 2060–70.

Broward County is also updating design requirements for public and private infrastructure, to account for recent and future changes in hydrologic conditions using integrated surface and groundwater hydrologic models informed by downscaled global climate data. The County's scenario-based planning employs the NCEP-Scripps Regional Spectral Model and integrates bias-corrected dynamically downscaled global-climate-model output from two GCMs, refined to a 10 km grid.

Model results provided the basis for establishment of "The Broward County Future Condition Average Groundwater Elevation Map" (Figures 1.1 and 1.2). This new map

Box 1.5 (cont.)

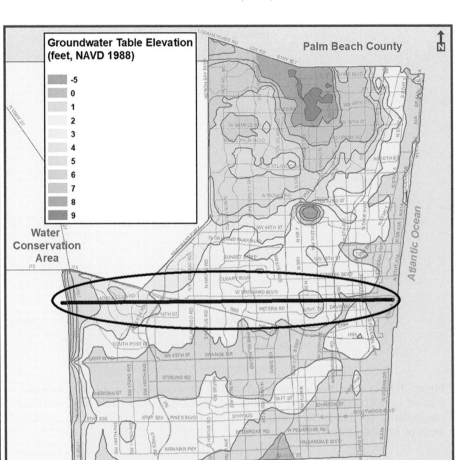

Figure 1.1 "Broward County's Future Condition Average Wet Season Ground-water Elevation Map," based on modeled outputs for the months of May through October over the period of 2060–9.

accounts for the modeled increase in groundwater table under conditions of sea level rise (26.6–33.9 inches) and increased rainfall (9.1 percent) projected for the period 2060–9. It applies to new development and major redevelopment within the County, to ensure that drainage and stormwater management infrastructure meet performance standards over the anticipated life of the investment. The County's future conditions

Continued

Box 1.5 (cont.)

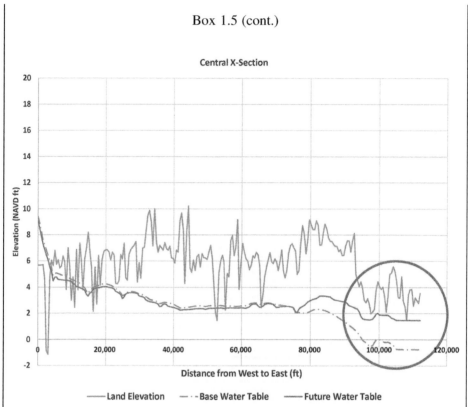

Figure 1.2 Cross section of modeled change in water table from west to east, relative to land surface, with a pronounced increase in water table from west to east, relative to land surface, and loss of storage capacity predicted for coastal areas of Broward County

100-year flood map is under development, incorporating the predicted shift in rainfall intensity in this same timeframe. Given the range of model results, a comparison of statistically and dynamically downscaled data sets is informing the range of values to be used in this analysis.

2

Global Climate Models

Global models of the Earth's climate system are the primary tools scientists use to understand the Earth's climate system. They yield critical insights into the components of the Earth's climate system, including the atmosphere, land, oceans, and biosphere, the processes at work within and between them, and how natural factors and human activities affect climate at the regional to global scale. This chapter summarizes the evolution of climate modeling and describes current global climate models and how they are being used to study the changing climate.

2.1 The Need for Climate Models

Computational models are often used by scientists and engineers to understand complex systems, as well as to understand the processes, interactions, and associated physics, chemistry, and biology affecting the world around us. In the case of the Earth's climate, such models are particularly important, as they allow scientists to construct a virtual "lab" where they can run experiments on an entire planet, whether studying the circulation of Jupiter's atmosphere, paleoclimate in the Earth's distant past, or how the climate today responds to the choices that humans make now and in the future.

The complex global climate models (GCMs) that simulate the Earth's climate have a variety of uses, including comparing them against observations to evaluate scientific understanding of individual components of the system and its processes; examining how these components respond to changes in both internal and external factors; determining how well we understand past and current changes in climate; and projecting how climate could change in the future. These models can incorporate both theoretical understanding and direct observations (e.g., observed changes in the output from the sun and documented changes in the emissions from human activities) to study the past and present response of the climate system to such changes, as well as providing the basis for projecting climate into the future.

Because of their unique ability to simulate the response of the Earth's climate system to human choices, and because their output takes the recognizable shape of maps, which enables scientists and decision makers to connect human choices to their resulting impacts at the regional scale, these models provide an important foundation for action on climate change. As a result, they often form the basis for analyses that range from understanding the science of climate change, to setting local, regional, national, or even international targets, to examining and comparing potential options for adaptation and mitigation.

2.2 The Evolution of Climate Modeling

The first mathematical models representing the Earth's energy balance and the processes affecting atmospheric radiative transfer were constructed the late 1800s (Edwards 2011). The best recognized of these early studies was by Swedish chemist Svante Arrhenius, who calculated by hand the effect of increasing concentrations of carbon dioxide on global temperature (Arrhenius 1896). Based on this study, using a simple climate model combining radiative transfer with a zero-dimensional energy balance model, Arrhenius correctly concluded that human emissions of carbon dioxide could warm the Earth. In his later book (Arrhenius 1906) he discussed the now well-recognized nonlinear relationship between an increase in the concentration of atmospheric CO_2 and the resulting change in global temperature, stating that "any doubling of the percentage of carbon dioxide in the air would raise the temperature of the Earth's surface by $4°$ (in degrees centigrade); and if the carbon dioxide were increased fourfold, the temperature would rise by $8°$."

Energy Balance Models (EBMs) such as Arrhenius used are one of the oldest types of climate models still in use today. EBMs estimate the changes in the Earth's climate from an analysis of the energy budget of the Earth. In their simplest form, zero-dimensional EBMs do not include any explicit spatial dimension; they simply determine a globally averaged temperature. The most commonly used form of EBMs today, however, is one-dimensional, where variations with latitude are accounted for by a parameterization of the dynamical effects of the Earth to model the transfer of energy across the planet (Budyko 1969; Sellers 1969). These models are generally used to evaluate the potential of different policies to reduce emissions of heat-trapping gases.

Radiative–convective models are the other commonly used simple type of climate model. These models allow energy from the absorption and reflection of solar radiation and emission of infrared to be exchanged both upward and downward between different layers in the atmosphere, averaged over the globe. They also consider the upward transport of heat by convection, which is especially important in the lower atmosphere, and are typically used to determine globally and seasonally averaged surface and atmospheric temperatures (Ramanathan and Coakley 1978). Radiative–convective models can also be used to study the effect of atmospheric gases and particles on surface temperature and on how temperature varies with altitude.

While simpler models can be useful tools to address particular questions, these models simplify and parameterize a great deal of important information that is necessary to fully understand the Earth's climate system. Therefore, to gain a fundamental understanding of the Earth's climate and the processes affecting it, a GCM is necessary. These complex, three-dimensional models are built on fundamental physical equations that include the conservation of energy, mass, and momentum, and how these quantities are exchanged among different components of the climate system. GCMs include the physics, the chemistry, and increasingly, the biology of various processes that make up the climate system. Using these fundamental relationships, GCMs are able to generate from first principles many important features that are evident in the Earth's climate system: the jet stream that circles the upper atmosphere; the Gulf Stream and other ocean currents that transport heat from the tropics to the poles; and even hurricanes in the Atlantic Ocean and typhoons in the Pacific Ocean, when the models are run at a fine enough spatial resolution. Biological processes are represented with increasing degrees of sophistication; these include different vegetation types and their growth, photosynthesis, roots, and leaves and the interactions of vegetation with nutrients in the soil and soil moisture. The atmosphere is made of trace amounts of gases and aerosol particles, in addition to nitrogen and oxygen. Models represent the most important of these trace gases and aerosols that effect the energy budget of the atmosphere by absorption or scattering, play a role in the formation and life cycle of atmospheric clouds. The science in these climate models is similar to what is used in weather forecast models, but focuses more on processes that affect the Earth system over periods of decades rather than hours.

GCMs divide the Earth, including the atmosphere, ocean, land surface, and cryosphere or ice, into grids of discrete "cells," which represent computational units (Figure 2.1). Unlike the simpler models described, physical processes internal to a GCM grid cell – such as convection – that occur on scales too small to be resolved directly by the model are parameterized at the cell level, while other functions govern the interface between cells.

The first three-dimensional computational climate model was developed by scientists at the US National Oceanic and Atmospheric Administration's (NOAA's) Geophysical Fluid Dynamics Laboratory (Manabe and Bryan 1969). These early global climate models were called General Circulation Models (using the same acronym, GCMs) as they were primarily concerned with simulating the general circulation of the atmosphere and used only a very simple representation of the ocean, assuming it was a well-mixed slab that provides moisture to the atmosphere.

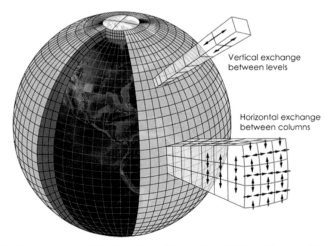

Figure 2.1 Computer models simulate climate by dividing the world into three-dimensional grid boxes, evaluating physical processes such as temperature, winds, and atmospheric concentrations of gases and particles at each grid point. At each model time step, a new state of the Earth's atmosphere and oceans is calculated and then used as the initial state for the next time step. By this method the model is "stepped forward" in time. The simulated climate can then be inferred by the "statistics" (averages, extremes, etc.) from multi-decadal simulations. Typical global climate model time steps are about thirty minutes. In practice, the time step also depends on the spatial resolution and, depending on the resolution and available computing power, models can be used to study changes in climate over years, decades, or even centuries and millennia.
Figure source: www.earthmagazine.org/article/todays-weather-forecast-good-strong-chance-improvement

As time went on, more processes and components were added to GCMs; and as computational power increased, so too did the spatial and temporal resolution of the models. For example, the model described by Manabe and Wetherald (1975) used an atmospheric horizontal resolution of about 500 km (roughly five degrees of latitude) with nine vertical layers, while today's models tend to have a horizontal resolution of one degree of latitude (~ 100 km) or finer and fifty to seventy or more vertical layers in the atmosphere, with a similar resolution in the ocean.

The historical spatial resolution of climate models has largely been constrained by the computational capabilities available at the time the models were being developed and run. It is therefore not surprising that the treatment of the processes affecting climate in these models has advanced greatly as the computational capabilities have increased, which in turn has increased understanding of relevant climate processes.

Figure 2.2 shows two different representations of the history of the development of climate models. Since the early GCMs of the 1960s and 1970s, GCMs have

A Climate Modeling Timeline
(When Various Components Became Commonly Used)

1890s	1960s		1970s	1990s	2000s	2010s
Radiative Transfer	Non-Linear Fluid Dynamics	Hydrological Cycle	Sea Ice and Land Surface	Atmospheric Chemistry	Aerosols and Vegetation	Biogeochemical Cycles and Carbon

Energy Balance Models   Atmosphere-Ocean General Circulation Models   Earth System Models

Figure 2.2 As scientific understanding of climate has evolved over the last 120 years, increasing amounts of physics, chemistry, and biology have been incorporated into calculations and, eventually, models. This figure shows when various processes and components of the climate system became regularly included in scientific understanding of global climate calculations and, over the second half of the century as computing resources became available, formalized in GCMs. Source: Hayhoe et al. (2017)

evolved to include more and more processes important to understanding the evolving climate system.

Box 2.1
Weather vs. Climate Models

While there are many similarities between models used for daily weather forecasts and the models used for climate projections, there are some important differences between the two. As discussed by IPCC (2013), to make accurate weather predictions, forecasters need highly detailed information about the current state of the atmosphere. The chaotic nature of the atmosphere means that even the tiniest error in the depiction of initial conditions typically leads to inaccurate forecasts beyond a week or so. In contrast, climate scientists do not attempt or claim to predict the detailed future evolution of the weather over coming seasons, years, or decades. Instead, climate models are used for understanding multi-decadal changes in the average statistics of temperature, precipitation, and other parameters that determine the climate system.

Weather forecasting is largely an initial-state problem, where understanding the initial conditions in detail is crucial. In contrast, climate is a boundary-value problem, where it is more important to know the initial climate conditions and the forcings that led to that climate representation. Climate model results over the long run, thirty years or more simulation time, produce similar climates with an increase in greenhouse gas emissions, even with small changes in how they are initialized, giving us confidence that this is a complex system that can be reliably modeled.

Many of today's GCMs are now called Earth System Models (ESMs), because they aim to encapsulate such a broad expanse of understanding of the physical, chemical, and biological processes involved in the climate system, their interactions, and the performance of the climate system as a whole. In addition to treating atmospheric and ocean dynamics and thermodynamics, ESMs include an interactive carbon cycle that accounts for biogeochemistry, and often simulate atmospheric chemistry and aerosols, and land-surface interactions including soil and vegetation, as well as land and sea ice.

2.3 Physical Processes in Global Climate Models

The processes that affect the Earth's climate operate on a range of different temporal and spatial scales, from chemical processes that occur in seconds at microscale to large-scale weather systems that extend over thousands of kilometers or ice-sheet processes that operate over millennia. These processes also interact with each other across different temporal and spatial scales. This means that even though a GCM may primarily be used for analyses and projections of climate over decades to centuries, scientists still need to incorporate processes on smaller spatial and shorter temporal scales into these models, because these processes interact with processes that occur on larger and longer scales.

The physical processes affecting the atmosphere that are incorporated into climate models can be divided into several categories. The first category consists of the physics that represents fundamental principles, such as the conservation of energy, momentum, and mass. The second category includes physical phenomena described by equations that are known to have closed-form solutions but that must, in practice, be approximated due to discretization of complex equations on computers. These equations include the transfer of radiation through the atmosphere and the equations that describe fluid motion. Complex partial differential equations are discretized through various computational techniques that maintain accuracy, mass conservation, and other highly desirable characteristics, while also being computationally efficient. The third category contains empirical equations, such as formulas for evaporation as a function of wind speed and vapor-pressure deficit. For the latter categories, modelers also often develop parameterizations that attempt to capture how a small-scale process acts over the scale of a GCM grid cell. For instance, the average cloudiness over a 100 km^2 grid box is not directly and easily related to the average humidity over the box. Nonetheless, as the average humidity increases, average cloudiness will also increase. Given uncertainties in the science, the approach used to develop a parameterization can depend on expert judgment. However, much of the large-scale behavior projected by climate models is robust, in that it does not depend significantly on the specifics of parameterization and spatial representation.

Figure 2.3 illustrates some of the major components of the Earth's climate that are considered in global climate modeling. Today's GCMs aim to simulate many processes, ranging from clouds to biogeochemistry to ice to marine ecosystems to large-scale weather events. A short description of some of the important components and processes that need to be represented in climate models are described below.

Clouds: Clouds play a significant role in controlling the energy and water budget of the planet. Representing clouds in all their details is one of the biggest challenges in climate modeling, as clouds range in size from less than a kilometer to giant organized convective systems. Given this range of cloud spatial scales, representation of clouds is sub-grid (smaller than grid resolution) of current generation of climate models and clouds are not explicitly resolved. Models of

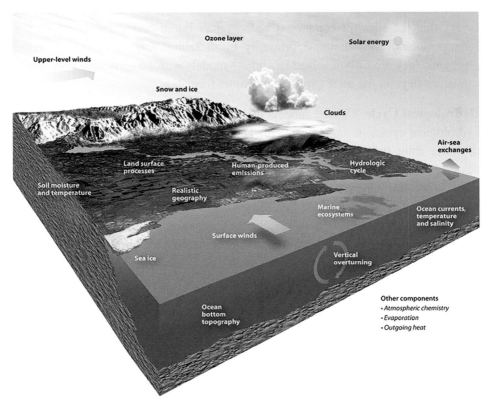

Figure 2.3 An illustration of some of the major components affecting the Earth's climate that are considered in a GCModel. (Graphic from NCAR; figure source: www2.ucar.edu/news/backgrounders/complexity-climate-modeling)

known physics of cloud formation and its lifecycle through precipitation are combined with parameters representing factors that are sub-grid to the model or poorly understood. The sub-grid phenomenon that is parameterized is the microscale process of cloud formation, seeded by submicron scale particles (e.g., Morrison and Gettelman 2008); the onset of precipitation; and the interactions between clouds and atmospheric particles that modifies their radiative properties. Climate models include two important types of cloud:

- Deep and Shallow Convective Clouds: These are small clouds that form low in the atmosphere, just above the atmospheric boundary layer, due to convective mixing in the atmosphere. Some of these grow into tall towers when the convection is strong and water vapor for condensation is available. These clouds are typically smaller than the grid size of a climate model and hence their formation and growth are extensive areas of research and analysis in climate models (Zhang and McFarlane 1995; Larson and Golaz 2005).
- Stratiform Clouds: These are low-level clouds, most common over marine environments and persisting for long periods of time, and are often seen as a very thin flat sheet of condensed water, either liquid or ice depending on the temperatures. They dissipate primarily by drizzling. Since they occupy large regions, they are fully resolved in climate models with no additional parameterizations beyond the microphysics of droplet formation.

Hydrological Cycle: Climate models include a detailed description of the Earth's hydrological cycles in the atmosphere. They represent the process of evaporation of water from the land, oceans, and large inland bodies of water; the clouds described above condense and precipitate water vapor and move large amounts of water in the atmosphere from one location to another; water vapor moves with the wind across the planet and deep into the atmosphere due to convection, advection, and mixing in the model's atmosphere. Climate models also represent the precipitation of snow and its melting in high latitudes and altitudes. Snow has a high albedo and reflects more solar radiation than does a snow-free surface, and is important for calculating the energy budget.

Atmospheric Circulation: GCMs in their first incarnation were essentially designed to represent the circulation of air around the planet. The circulation of air transports heat and water vapor, mixes surface air with upper atmosphere, and essentially determines the weather on any given day, by its interaction with ocean and land, the climate over longer time scales. The main arm of the atmospheric circulation moves warm air up from the tropics up and toward the poles, where it sinks to the surface of the Earth and brings the cold air back toward the tropics. There are several other persistent flow systems imposed on this overall flow pattern, some planetary scale waves at high latitudes, jet streams, westerlies, and easterlies that are represented in these models. Many climate models are designed using numerical schemes that divide the atmospheric flow into wave motions of several wave lengths. Often the model designation used by a particular model refers to the shortest atmospheric wave resolved. For example, a T85 model truncates the atmospheric flow at wave number 85, which has an approximate size of 100 km in the horizontal. Atmospheric waves shorter than this wavelength are not represented in the models, and their effect on atmospheric flow is parameterized. In the vertical, atmosphere is represented as various layers extending from the surface to the top of the atmosphere. These layers are: the atmospheric boundary layer (the lower 2 km of the atmosphere), troposphere (the lower 15–18 km of the atmosphere), stratosphere (from the top of the troposphere to about 50 km) and in some models the mesosphere (from 50 km to 80 km).

Land Surface Representation: The representation of land, properties of the land for conducting and storing heat, infiltration of water through the soil to ground water table, and topography of the land are all key elements included in the model and define the coupling of the land to atmosphere. Surface runoff of water through gravity flow from land to oceans is included using simple representations of the rivers and streams. On Earth, the heating and cooling of the land, which drives the diurnal circulation patterns of temperature and atmospheric motion, are represented in the model. The land includes several layers of soil and represents the storage of heat in the lower layers of soil, which changes over longer time periods (seasonal and annual scales) as compared to the surface layer of the soil, which responds to solar heating each day. Precipitation, and its effect on soil moisture, is represented, making it possible to model the coupling of soil moisture with atmosphere through evaporation of moisture from the soil and the heating of the atmosphere through transfer of infrared radiation from the soil to the atmosphere. Topography at the spatial resolution of the model is included, as it has important effects of atmospheric flow, cloud formation, and precipitation.

Biogeochemistry and Atmospheric Chemistry: Climate models increasingly include complex sub-models of biogeochemistry and atmospheric chemistry. The biogeochemical cycles of carbon, nitrogen, and phosphorous are now included in many state-of-the art models. Carbon-cycle models include the entire life cycle of carbon in the atmosphere: its transport; vegetation as a sink during the growing season and release back into the atmosphere during the senescence in fall/winter; transfer of carbon to the soil for long-term storage; and carbon sink to the oceans. The nitrogen cycle is represented in terms of its sources from anthropogenic and natural emissions and removal through precipitation. Other trace atmospheric gases, such as nitrous oxide and methane, which are important heat-trapping gases, are included in many GCMs. Models that represent methane include other key atmospheric gases, such as ozone and hydrocarbons, as they are required to represent the processes that lead to atmospheric loss of methane due to chemical reactions.

Incoming Solar Energy and Outgoing Heat Energy. The GCMs were initially conceived as EBMs as discussed earlier and thus considerable efforts were made to represent the transfer of shortwave energy from the sun through the atmosphere to the surface of the Earth. The return of most of this energy back into the space from the top of the model as it is reflected back from the clouds, land, ocean, and atmospheric gases and particles is also modeled. The remaining energy absorbed by the surface, aerosols, and atmospheric gases released back into the atmosphere as infrared radiation is represented. The input from sun includes the representation of variability of energy from the sun in the form of sunspot activity, seasons, and location on the Earth and the time of the day.

Glaciers, Land, and Sea Ice: GCMs include models of land-based ice sheets and ice cover over the ocean to various degrees of complexity. Ice-sheet models have been recently introduced in various climate models and made available to the next IPCC assessment and the CMIP6 model archive. Ice-covered oceans and ice sheets have high albedo and reflect the shortwave radiation reaching the surface, and act to cool the lower atmosphere near the poles. Melting sea ice could expose the darker oceans in the polar region, and the dark regions absorb more of the incoming energy, which could lead to warmer oceans and more ice melt. Representing the ice and ice sheets accurately is thus critical for climate models and one topic that is of large scientific interest at present.

Ocean Currents, Temperature, and Salinity. Climate models have evolved from representing oceans as simple slabs that primarily acts as a heat sink to balance the Earth's energy budget to models that now represent the circulation of the ocean currents. Ocean models calculate the temperature of the ocean water at various depths of the ocean, mixing within the different layers of the ocean, ocean salinity, and acidity. Modeling these variables is essential for representing phenomena such as the North Atlantic's overturning circulation, which helps keep Europe relatively warm for its latitude. Representing the temperature gradient through the depth of the ocean, overturning circulations, and marine biogeochemistry are also essential for representation of carbon cycle over timescales of decades and longer. These processes are now present in the more advanced climate models.

2.4 Advances in Climate Modeling and Model Resolution

Representations of Earth system processes have significantly improved in recent years as a result of new observations and process analyses. Notable advances have been made in representing effects of aerosols on radiative transfer and various ways that aerosols and clouds interact, capturing biosphere representation in the models and treating the cryosphere (ice). A representation of the carbon cycle has been added to many models; those that have a carbon cycle are called ESMs. Also, a high-resolution stratosphere is now included in many models so that stratospheric ozone and dynamical interactions can be better treated. Other ongoing process developments in climate models include enhancing the representation of nitrogen effects on the carbon cycle. Looking to the future, current efforts are focusing on improving model treatment of the processes affecting ice-sheet dynamics, land-surface hydrology, and the effects of agriculture and urban environments. As new processes or treatments are added to the models, they are also evaluated and tested relative to available observations.

In addition to including additional processes in the models and improving the treatment of existing processes, the total number of climate models and the average horizontal spatial resolution of the models have improved over time. There are now over forty GCMs that have been created and are run by research groups at government laboratories or research facilities around the world. While the basic structure of these models is comparable, they all differ in their details: The Earth's climate system is incredibly complicated. These models must portray the physical interactions among the atmosphere, the oceans, land surfaces, and sea ice with respect to a multitude of processes operating on many different space and time scales. Different models make different choices as to which elements of the physics to emphasize, for instance by how finely the vertical structure of the atmosphere is subdivided. Different models also have different portrayals of elements of the climate system that are more challenging to model, such as the treatment of clouds, aerosols, or the carbon cycle.

The utility of having a number of independent climate models is in that it gives us an opportunity to evaluate models that use different assumptions, physics parameterization and computational frameworks relative to one another. These types of model intercomparisons help us understand the response of the models to external forcing, perturbations, and variations in their projections as a result of the model's ability to capture complex coupling between the various components of the model and its expression as natural variability. International model intercomparisons are designed to extract these features from the models, to identify model sensitivities and to generate enough ensembles to provide a reasonable sample size to address projection uncertainties that are discussed in detail in Chapter 7.

Atmosphere and ocean processes represent a continuum of scales spanning at least ten orders of magnitude. Figure 2.4 shows the interplay between resolution and the processes that can be represented across various temporal and spatial scales

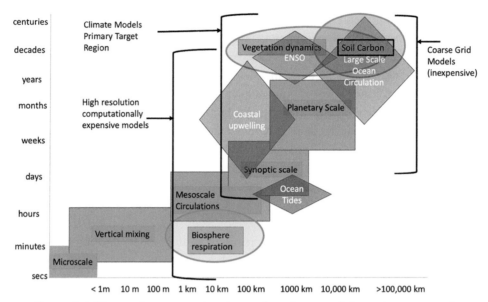

Figure 2.4 Time and space overlapping scales of some of the major ocean and Earth processes important to climate analyses. The chart defines the required resolution and sampling interval needed to capture the features of interest. The *x*-axis represents the spatial scale and the *y*-axis represents the timescale of the process.

for some of the major ocean and Earth processes important to climate analyses. Figure 2.4 also shows the approximate range of spatial and temporal scales that can be accounted for with different resolution climate models run at 2-, 1-, or 0.25-degree atmospheric resolution.

Higher spatial resolutions in climate models improves the models' ability to resolve a number physical biological and oceanic processes. For example, studies (e.g., Zobel et al. 2018) have shown that local precipitation patterns are much better represented at the finer resolution. Figure 2.5 illustrates how model ability to resolve precipitation improves significantly from using a horizontal grid spacing of 100 km (similar to most current GCMs) to a grid spacing of 25 km (the finest resolution being used currently). Increasing spatial resolution also improves models' ability to simulate large-scale circulation features such as the jet stream, which has important effects on the treatment of atmospheric rivers and associated heavy precipitation. As computers become more powerful, even higher resolution will be possible, perhaps down to the few-kilometer level necessary to resolve clouds.

A significant constraint on running a complex GCM at high resolution is the computational cost. Models usually run on some of the most powerful supercomputers in the world, the costs include not only the runtime, but also

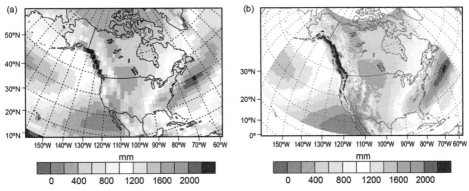

Figure 2.5 Current GCMs typically operate at coarser horizontal spatial scales on the order of 50–300 km, while extremely high resolution runs with climate models are now reaching even finer resolutions, on the order of 10–50 km. This figure compares annual average precipitation (in millimeters) for the historical period 1979–2008 using (a) a resolution of 250 km with (b) a resolution of 25 km to illustrate the importance of spatial scale in resolving key topographical features, particularly along the coasts and in mountainous areas. In this case, both simulations are by GFDL HiRAM (the Geophysical Fluid Dynamics Laboratory High Resolution Atmospheric Model), an experimental high-resolution model.
Source: Hayhoe et al. (2017) adapted from Dixon et al. (2016)

the infrastructure necessary for storing the large datasets produced by running the model, currently on the order of petabytes. Over the last decade, primarily thanks to improvements in computational ability, there has been a rapid increase in the horizontal and vertical resolution of climate models. The atmosphere, oceans, and land processes are all being treated with ever finer resolutions, currently as fine as 0.25° (~25 km/15 miles) horizontal resolution in the atmosphere and 0.1° (~10 km/ 6 miles) in the ocean. The refined and often more sophisticated grids being used in the ocean and atmosphere models are aimed at making optimal use of today's computer architectures.

2.5 Evaluating Climate Models Using Past Climate

Scientists have been collecting data on climate for many decades, including Antarctic ice cores, tree rings, coral, and boreholes in soils and ocean sediments, to determine what the Earth's climate was like in the past. From this research they have discovered details about past human activity, temperature changes in our oceans, periods of extreme drought, and much more. Despite their complexity, GCMs are now being run on today's supercomputers for extended simulated time periods, sometimes for thousands of model years in the past and for projections of several hundred model years into the future. Climate models can be evaluated using data from these past periods to understand those past changes and the processes that can lead to climate change. Although significant uncertainties remain in the reconstructions of past climate variables from proxy records and forcings and in understanding the forcing driving those past changes in climate, paleoclimate information from the Mid Holocene, Last Glacial Maximum, and Last Millennium have been useful in testing GCMs and their ability to simulate the magnitude and large-scale patterns of past changes. These simulations also establish a baseline for studying future change and build confidence in the projections of how the climate will function under very different conditions from today.

The first type of simulation done by any global climate model is a "control simulation," where initial conditions are typically set to match those experienced in a time before the Industrial Revolution, and the model is run for several thousand years to allow the different components of the climate system, particularly the ocean, to come into balance with each other. Once a control simulation has reached equilibrium, it can then be used to initialize a transient or time-sensitive historical simulation that aims to reproduce the observed influence of human and natural factors on climate over that time. Historical simulations are initialized in a past year, typically 1850. This is considered the preindustrial revolution period and before extensive use of fossil fuels and the rapid increase of CO_2 in the atmosphere. They are then run with prescribed forcings (scientists' best estimate of what really happened in that year) for each year until 2005 (in CMIP5 simulations) or 2014 (in CMIP6 simulations). The large archive of GCM simulations results and the process used to create this archive for CMIP5 and CMIP6 is described in detail in Section 2.6. These prescribed forcings are both natural (changes in solar irradiance, volcanic eruptions) and human-related (emissions and concentrations of greenhouse gases and aerosols, plus effects from land use/cover change) (IPCC 2013).

Unlike weather forecasts, historical climate simulations are not periodically adjusted with updated information about the state of the climate to improve the forecast – they are initialized for a historical year, then constrained only by the prescribed forcing from then on. Owing to uncertainties in model formulation and observations of the initial state of the climate system (which, in 1850, were quite uncertain), any individual simulation represents only one of the possible pathways the climate system might follow. This is because of the sensitivity of natural variability in the climate system, including the timing of quasi-periodic (and quasi-chaotic) natural cycles, such as El Nino and the Pacific Decadal Oscillation, to initial conditions. As a result, historical simulations are not designed (or expected) to reproduce the observed sequence of weather and climate events during the twentieth century. In that sense, they are best viewed as an alternative planet, with the same factors influencing climate as ours, but with its own patterns of natural variability and weather. However, these simulations *are* designed to reproduce observed multi-decadal climate statistics, such as averages. And they can be used to study specific events in the observational record.

GCMs are also extensively tested relative to modern observations, and many studies have demonstrated they are able to reproduce the key features found in the climate of the past century and contribute to scientific understanding of observed trends. One example of a recent event that GCM simulations were useful in understanding was the observed increase in global temperature for the fifteen years following the 1997–8 ENSO event, which was smaller than the underlying long-term increasing trend over thirty-year climate time scales (Fyfe et al. 2016). The increase in the heat content of the entire climate system continued apace (Benestad 2017), but the proportion of heat going into the atmosphere was smaller than average. Variations in the rate of warming over relatively short timescales of a decade or two are expected due to long-term internal variability in the climate system, including changes in the exchange of heat between the atmosphere and ocean, or short-term changes in climate forcings such as aerosols or solar irradiance. However, the temporary slowing in the rate of global temperature increase over this time was wrongly used by some to cast doubt on the accuracy of climate projections from CMIP5 models, since the measured rate of warming in all surface and tropospheric temperature datasets from 2000 to 2014 was less than expected given the results of the CMIP3 and CMIP5 historical climate simulations (Fyfe et al. 2016; Santer et al. 2017).

In fact, GCM simulations helped scientists determine that it is very likely that the early 2000s slowdown was caused by a combination of short-term variations in forcing and the exchange of heat between the ocean the atmosphere, the same factors that were known in advance, though the relative contribution of each is still an area of active research (e.g., Trenberth 2015; Fyfe et al. 2016; Meehl et al. 2016). The temporary slowing soon ended, though, with 2016 currently being the warmest year on record, and 2015 and 2017 the second and third, respectively. These more recent warm years brought the observed temperature trends the long-term thirty-year trends into strong agreement with the CMIP5 model results, as shown in Figure 2.6. A second important point illustrated by Figure 2.6 is the broad overall agreement between observations and models on a timescale of a century, which is robust to the shorter-term variations in trends in the past decade or so. Continued global warming and the frequent setting of new high global mean temperature records or near records is consistent with expectations based on model projections of continued human forcing driving long-term increases in global temperature.

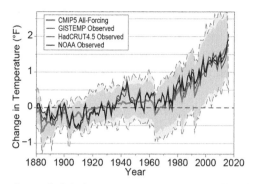

Figure 2.6 Comparison of global mean temperature anomalies (°F) from observations (through 2016) and the CMIP5 multi-model ensemble (through 2016), using
the reference period 1901–1960. The CMIP5 multi-model ensemble (dark-gray
range) is constructed from blended surface temperature (ocean regions) and
surface air temperature (land regions) data from the models, masked where
observations are not available in the GISTEMP dataset. The importance of using
blended model data is shown in Richardson et al. (2016). The thick solid-gray
curve is the model ensemble mean, formed from the ensemble across thirty-six
models of the individual model ensemble means. The shaded region shows the
+/− two standard deviation range of the individual ensemble member annual
means from the thirty-six CMIP5 models. The dashed lines show the range from
maximum to minimum values for each year among these ensemble members. The
sources for the three observational indices are: HadCRUT4.5 (red); NOAA
(black); and GISTEMP (blue). (NOAA and HadCRUT4 downloaded on Feb. 15,
2017; GISTEMP downloaded on Feb. 10, 2017).
Sources: USGCRP (2017) and adapted from Knutson et al. (2016)

What will happen in the future? Future scenarios are discussed in more detail
in Chapter 7, but for context, projected changes in globally averaged temperature
for a range of future pathways that vary from assuming strong continued
dependence on fossil fuels to rapid reduction and eventual elimination of heat-
trapping gas emissions before the end of the century were estimated by the
CMIP5 models. Global surface temperature increases for the end of the twenty-
first century are *very likely* to exceed 1.5 °C relative to the 1850–1900 average
for all projections, with the exception of the lowest part of the uncertainty range
for the lowest scenarios (IPCC 2013; Peters et al. 2013; Knutti et al. 2016;
Schellnhuber et al. 2016).

2.6 Archives of GCM Simulations

GCMs are formally compared with each other and with observations as part of a
periodic assessment and archiving of GCM simulations known as the Coupled Model

Intercomparison Project (CMIP), which occurs under the auspices of the World Climate Research Programme. The aim of CMIP is to better understand past, present, and future climate changes arising from natural, unforced variability or in response to changes in radiative forcing in a multi-model context. This understanding includes assessments of model performance during the historical period and quantifications of the causes of the spread or uncertainty in future projections.

The first source of uncertainty in GCM simulations is parametric, introduced when GCMs require parameterizations to simulate processes that occur on spatial or temporal scales smaller than they can resolve, such as precipitation and cloud formation. The second type of modeling uncertainty is structural. This addresses the extent to which the global model includes – and accurately represents – important physical processes occurring on scales they can resolve. Structural uncertainty can arise because a process is not yet recognized, or because it is known but not yet understood well enough to be modeled accurately. As discussed previously, improving structural uncertainty is a key aspect of model improvement, as more processes and factors are added to GCMs and evaluated with each successive generation. Structural and parametric uncertainty in global model simulations, as well as other important sources of uncertainty including natural variability and human choices, are discussed in more detail in Chapter 7.

To further characterize these sources of uncertainty, CMIP also uses idealized experiments to increase understanding of the model responses. In addition to the long-time-scale responses discussed, numerical experiments are performed to investigate the predictability of the climate system on various time and space scales and making predictions from observed climate states. Initializing the models with the observed climate state, instead of 1850 for example, should provide model outputs that are similar to our current observed climate state at a much higher time scales than as an average over a climate epoch.

An important goal of CMIP is to make the multi-model output publicly available in a standardized format. To that end, CMIP periodically archives output from GCMs. Many of the archived simulations are idealized experiments or reconstructions of the distant past, useful mainly to scientists. The simulations most relevant to real-world applications are: (1) the historical total-forcing simulations that represent each modeling group's best attempt to reproduce observed natural and human influences on climate from 1850 to the present day; and (2) the future projections where simulations corresponding to a range of future scenarios of human activities paint different pictures of what the world might look like, depending largely on the choices humans make today and in the near future (see Chapter 7 for more details).

Box 2.2
Can We Trust Climate Models?

Scientists use a wide range of observational and computational tools to understand the complexity of the Earth's climate system and to study how that system responds to external forcings, including human activities. Climate models are just one of those tools – yet often they are singled out as if they were the sole basis for the state of understanding climate change. Some express distrust of climate models – suggesting models are "no good" or "too uncertain." But is that distrust warranted? The answer is: No.

Climate models have proven remarkably accurate in forecasting and evaluating the climate change we have experienced to date. It is true that scientific understanding of physics, chemistry, and biology is incomplete, and the models necessarily incorporate those uncertainties. In some cases, model analyses have been overly conservative: for example, in projecting how quickly Arctic sea ice would decline. Models also tend to underestimate the observed increase in large precipitation events. In such cases, the observations provide clues on how the representation of physical processes can be further improved in the models. These models are the only tools we have with significant predictive capacity – and although not perfect, they are very useful tools and provide us with substantial insights into future climate.

Because models differ in their representation of certain processes, scientists make use of these differences by examining ensembles of model runs and suites of models in climate assessments. However, despite the tremendous improvements in the climate modeling capabilities since the first GCM-based study of climate change was done over forty years ago (Manabe and Wetherald 1975), it is interesting to note that the most significant simulated response of the climate system to human activities – a substantial increase in average global temperature – continues to be about the same as forecast by the original GCMs decades ago.

CMIP phase three (CMIP3) ran from 2005 to 2006 (Meehl et al. 2007) and provided output for more than a dozen climate models for the historical total-forcing simulations (to 2000) and a range of future simulations known as SRES (Special Report on Emission Scenarios; Nakicenovic et al. 2000). The simulations archived by this dataset were used in the fourth IPCC assessment report (AR4; IPCC 2007) and in the second and third US National Climate Assessments.

CMIP5, which ran from 2008 to 2012, created an even more extensive archive of simulations based on output from over fifty GCMs. The average horizontal spatial resolution in CMIP5 was much higher than CMIP3, ranging from about 50 to 300 km (30–200 miles) vertical spatial resolution of hundreds of meters in the troposphere or lower atmosphere. It also conducted a more detailed comparison of climate models with each other and with observations (Taylor et al. 2012). It provided output for the historical total-forcing simulations (to 2005) and a range of future scenarios known as RCPs (Representative Concentration Pathways; Moss et al. 2010). Simulations from CMIP5 were used in the fifth IPCC assessment report (AR5) and in the fourth US National Climate Assessment (USGCRP 2017).

It is important to note that, increasingly, modeling groups provide output from ensembles: multiple simulations of the same scenario, each one run with slightly different initial conditions. Using ensembles from a single model or multiple models increases the statistical sample and helps both scientists and users to encompass the range of variability in weather statistics, particularly over shorter time scales; using a multi-model ensemble additionally helps to account for uncertainty in model formulations. As mentioned in Section 2.3 and in more detail in Chapter 7, it is well recognized in the science community that ensembles allow for better understanding of the role of natural variability in studies to understand the potential magnitude of climate change.

The newest international model intercomparison, CMIP6, began in 2018. It currently provides simulations from twenty GCMs and is expected to reach approximately fifty models when all the simulations are archived. Major differences from CMIP5 include even more GCMs; an additional increase in the average spatial resolution of the models; and, as discussed in Section 2.3, a significant expansion in the types of physical processes included in the models. As with earlier phases, most CMIP6 simulations are relevant only to scientific analyses; however, as with earlier phases, CMIP6 also archives historical total-forcing simulations (to 2014) and a range of future scenarios known as SSPs (Shared Socioeconomic Pathways) that correspond to the RCPs used in CMIP5 in terms of their influence on climate, but include a host of other socioeconomic data to accompany them that allows for better quantification of both drivers and impacts of climate today.

All three archives, CMIP3, CMIP5, and CMIP6, are maintained and made available through the Program for Climate Model Diagnosis and Intercomparison (PCMDI) at Lawrence Livermore National Laboratory (https://pcmdi.llnl.gov).

Box 2.3
Weighting Global Climate Model Results: A Case Study

Anyone who works with GCMs knows that some models are better than others – but the question is where, and at what? Some models are more successful than others at replicating observed climate and trends over the past century; some, at simulating the large-scale dynamical features responsible for creating or affecting the average climate conditions over a certain region, such as the Arctic or the Caribbean (e.g., Wang et al.

Continued

Box 2.3 (cont.)

2007; 2014a; Ryu and Hayhoe 2013); others, at simulating past climates with very different states than present day (Braconnot et al. 2012).

These differences have led to discussions of whether the different climate models used in national and international assessments of climate change should be weighted, based on how well the models represent observations. Until the fourth US National Climate Assessment (NCA4), previous assessments used a simple averaging of the multi-model ensemble. That approach implicitly assumes each climate model is independent of the others and of equal ability. Neither of these assumptions, however, is entirely valid. Some models share many components with other models in the CMIP5 archive, whereas others have been developed largely in isolation (Knutti et al. 2013; Sanderson et al. 2015). Also, some models are better than others at simulating certain aspects or regions of the climate system but ranking a specific model's performance often depends on the variable or metric being considered in the analysis. Some models perform better than others for certain regions or variables, but a completely different result might be obtained if the models are applied to a different variable or region. Therefore, ranking a specific model's performance often depends on the variable, metric, or region under consideration for the analysis. However, all simulations of future climate do agree that both global and regional temperatures will increase over this century in response to increasing emissions of greenhouse gases from human activities.

In earlier studies on weighting, model weights were often based on historical performance for a limited number of variables; however, it turns out such rankings, based on historical performance, may not improve future projections (Knutti and Sedláček 2013). For example, ranking GCMs based on their average biases in temperature gives a very different result from when the same models are ranked based on their ability to simulate observed temperature trends (Jun et al. 2008; Giorgi and Coppola 2010). If GCMs are weighted in a way that does not accurately capture the true uncertainty in regional change, the result can be less robust than an equally weighted mean (Weigel et al. 2010). The intent of weighting is to increase the robustness of projections, but by giving lesser weight to outliers a weighting scheme could increase the risk of underestimating the range of uncertainty.

To address these challenges, NCA4 used a modified form of model weighting to refine future climate change projections. In NCA4, model independence and selected global and North American model-quality metrics were both considered in the weighting parameters (Sanderson et al., 2017, building upon the earlier study by Knutti et al. 2017). The weighting approach is unique, as it takes into account the extent to which individual climate models share code as well as their relative abilities in simulating North American climate, the focus of the study (again, a different set of weights may be obtained for a different region). Understanding of model history, together with the fingerprints of particular model biases, was used to identify model

Box 2.3 (cont.)

pairs that are not independent. Thus, this approach considers both skill in the climatological performance of models over North America, the region of interest, as well as the interdependency of models arising from common parameterizations or tuning practices. The weights, once computed, can be used to simply compute weighted mean and significance information from an ensemble containing multiple initial condition members from codependent models of varying skill.

Evaluation of this approach shows improved performance of the weighted ensemble over the Arctic, a region where model-based trends often differ from observations, but little change in global-scale temperature response and in other regions where modeled and observed trends are similar. The choice of metric used to evaluate models has very little effect on the independence weighting, and some moderate influence on the skill weighting if only a small number of variables are used to assess model quality. Because a large number of variables are combined to produce a comprehensive "skill metric," the metric is not highly sensitive to any single variable.

3

Assessing Climate-Change Impacts at the Regional Scale

Although climate change is a global issue, its impacts are experienced primarily at the local to regional scale. This chapter describes important aspects of regional climate and how climate projections can be used to assess climate impacts at the regional to local scale. It summarizes projections and sources of information on changes in continental-scale annual and seasonal temperature and precipitation, climate and weather extremes, and sea level rise projections.

3.1 Climate Projections for Regional Assessments

Assessments of regional climate change impacts provide a way to evaluate the potential effects of climate change on issues relevant to a given region and/or sector. Common topics for regional assessments include region-specific observed and projected changes in aspects of climate relevant to that area, such as changes in average temperature or precipitation, extreme weather, sea level rise, and more.

Assessments also often include analysis of how these changes will affect one or more important sectors for that region, such as:

- Agriculture, including crop yield, livestock production, and shifts in the geo-graphical range or timing of pests
- Ecosystems and ecosystem services, such as the geographic or phenological shifts in a specific invasive or at-risk plant, animal, or bird species, and impacts on the functionality of the regional ecosystem as a whole
- Human health and welfare, including heat-related illness and death, air pollution, extreme events, nutrition, social-justice issues, and more
- Infrastructure, including the vulnerability of transportation, the built environ-ment, communities, public services, and observed and projected future changes

- Natural resources, such as land use, and the supply of and demand for food, water, and/or energy
- Other climate-sensitive sectors, from tourism and national security to fisheries and coastal resources.

These assessments can support the development of robust strategies that increase the resilience of both human and natural systems to climate change. They can also provide important guidance regarding how to prioritize, encourage, and support adaptation and resilience in vulnerable areas.

Region-specific climate projections form the basis of most regional climate assessments that address future impacts. Projections can be used *qualitatively*, to inform the direction of future change and enable an approximate idea of the magnitude. For example, is a given region warming faster or slower than the global average? Are extreme heat and/or extreme precipitation becoming more common? Is the region likely to become drier or wetter, and how will this affect soil moisture? Which season(s) will see the greatest change? To answer these questions for a relatively large region or nation on the order of roughly two million square kilometers (or one million square miles or more), give or take, such as the size of the US Midwest or the coastal Mediterranean area, GCM output can be sufficient and relevant information can often be obtained from national assessments, such as for Australia (CSIRO and Bureau of Meteorology 2015), Canada (Bush and Lemmen 2019), the European Union (Füssel et al. 2017), the United Kingdom (UKCP18 2018) the United States (USGCRP 2018) or even international assessments or databases such as the IPCC and the CMIP archives (see Chapters 1 and 2). This approach formed the basis of the First US National Climate Assessment (USGCRP 2000) as well as several regional assessments conducted by the state of California (Bedsworth et al. 2018), the Gulf Coast (Watson et al. 2015) and the Great Lakes (Kling et al. 2003). The second half of this chapter, Section 3.2, summarizes and provides sources for qualitative projections for each major world region.

Box 3.1
Agricultural Impacts

Impacts of changing climate on crop yields for corn are estimated to be between -10 percent and -40 percent for most of the midwestern states of the United States under RCP4.5 scenario by mid-century. Fertilization effects of excess CO_2 in the atmosphere are expected to compensate for some of the yield loss due to changes in temperature and precipitation by between -5 and -30 percent (Jin et al. 2017). Knox et al. (2012) performed a meta-analysis of published literature on crop yield impacts from climate change for Africa and South Asia and estimated a decrease of -8 percent for all crops by 2050 for these regions.

The second way projections are used in regional climate assessments is *quantitatively*, to evaluate likely changes in the risk of specific events, such as the number of days per year over a given temperature or precipitation threshold, or the return period of a historic event such as a heat wave or flood. Quantitative climate projections can also be used as input to impact models that can calculate how climate change might affect specific aspects of the regional economy and its resources, such as crop yields, species abundance, energy demand, water supply and snowpack, infrastructure lifetime, disease risk, and more. For example, the impact of changing climate on the European Economy in four economic sectors of agriculture, floods, coastal regions and tourism would lead to decrease in household welfare between 0.2 and 1.0 percent, if conditions projected for 2080 are imposed on current economy (Ciscar et al. 2011). An assessment of the GDP impact from climate change for over 130 countries (Kompas et al. 2018) shows that economic impact in terms of GDP loss at more than -10 percent per year for countries in Southeast Asia and India in the long-term. The fourth US National Climate Assessment (USGCRP 2018) estimates a GDP loss for the United States in the range of 3–10 percent in the long-term.

In addition to providing projections that accurately reproduce observed climate and the influence of topography and land use on fine-scale variations in climate across the landscape, high-resolution climate projections can also be key to resolving changes in climate extremes, at the tails of the distribution of a given climate variable, and changes that are significantly moderated, or even driven, by topography or geographical features that are not fully resolved by GCMs. Most regional assessments conducted over the last two decades include high-resolution climate projections, such as the fourth US National Climate Assessment (NCA4; USGCRP 2018b), which used statistical downscaling using the Localized Constructed Analogues (LOCA), (Pierce et al. 2014) model projections that were customized and developed specifically for that assessment. Some other climate assessments, such as the state of California (Bedsworth et al. 2018) and the city of Chicago (Hayhoe et al. 2010) have also used downscaled projections generated specifically for that particular assessment. The European Union (EU) has developed customized downscaled products using both dynamic and statistical downscaling for their recent assessment (Füssel et al. 2017).

To generate the high-resolution climate projections required for regional assessments and as input to subsequent impact modeling by crop and soil scientists, ecologists, civil and hydraulic engineers, epidemiologists, urban planners, water managers and more, a broad suite of methods – collectively known as bias correction and downscaling techniques – have been developed. These methods convert GCM output into the spatial (and sometimes even temporal) scales required to answer the needs of local and regional stakeholders and decision makers. They also ensure the GCM output is more representative of local- or regional-scale climate behavior and changes, both in terms of spatial resolution and capturing key phenomena that occur at these smaller spatial scales.

Bias correction and downscaling methods are often combined with analysis that additionally translate the raw high-resolution climate variables they produce, such as daily temperature, precipitation, relative humidity, or solar radiation, into impact-relevant metrics or indicators, such as days below freezing or heavy precipitation events, or into the formats required for input into impact modeling, including Geographical Information System (GIS) layers. This process does not have to be conducted separately for each sector; typically, the creation of high-resolution climate projections for a given location or region generates climate variables that can be used for a variety of purposes, from calculating how often a historical heat wave event will recur to using them as input to a hydrological model that forecasts local or regional flooding under given precipitation conditions. Figure 3.1 illustrates the spatial progression of climate projections from global model output (left) to gridded regional projections (center), to ultra-high-resolution local scale gridded or station-level projections (right).

Using high-resolution projections is important, as there can be significant differences in climate projections at large regional scales (Figure 3.1, middle) compared to global average changes (Figure 3.1, left). Local changes (Figure 3.1, right) can additionally differ from regional projections and even occasionally be of the opposite direction to that of other locations within the same region. Thus, for impact studies, the use of regional average values may not capture anticipated

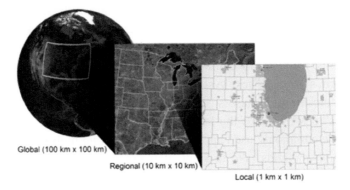

Global (100 km x 100 km)
Regional (10 km x 10 km)
Local (1 km x 1 km)

Figure 3.1 GCMs are global in scope and have grid cells that are typically on the order of hundreds of km per side. Regional climate models operate on a finer spatial resolution on the order of tens of km per side. Ultra-high-resolution regional climate model simulations or empirical–statistical downscaling to observational gridded data typically occurs at resolutions of a few kilometers or less, or even individual stations. These are considered to be local or (if applied to cities) urban scales.

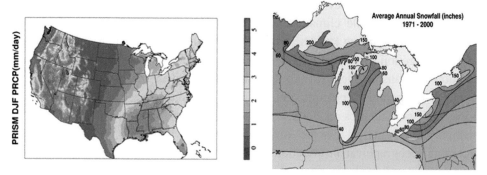

Figure 3.2 (Left) Observed precipitation over the contiguous United States for winter months (December, January, and February) averaged for the years 2000-2009. The precipitation pattern shows the influence of terrain changes over California and the Rockies, synoptic scale systems over the Southeast and Pacific Northwest and influence of the Great Lakes on precipitation in the upper Midwest. (right) Annual average snowfall is highest along the eastern coast of Lakes Michigan, Superior, Huron, Ontario, and Erie in the upper Midwestern US and Canada.

changes for the system of interest. What gives rise to this heterogeneity in responses across spatial scales? Key drivers include terrain, land use and landcover differences, land–ocean interfaces, inland hydrological characteristics, and persistent regional-scale phenomena that are often driven by these features.

The importance of regional drivers in determining precipitation and precipitation change over a region is illustrated by Figure 3.2. Average annual precipitation along the western coast of the United States is driven primarily by terrain: moist air off the Pacific Ocean drops precipitation on the west side of the mountain range as it rises; by the time it reaches the eastern side of the mountains, the air is warm and dry (Figure 3.2, left panel; Cascade Range along the Northwest, Sierra Nevada in California). In the Midwestern US, in contrast, precipitation is primarily driven by large-scale storm systems. These systems are moderated by the Great Lakes, which provide moisture to air masses as they move over them, creating strong "microclimates" downwind of the lakes (Figure 3.2, right panel).

There are two commonly used approaches to generating high-resolution climate projections. Both methods take account of regional-scale topography and other important regional features, but they accomplish this in very different ways. Dynamic downscaling uses high-resolution regional climate models (RCMs) that resolve features such as mountains and lakes at the scale of their grid, typically 10 km per side. Empirical–statistical downscaling models (ESDMs) combine GCM output with real-world observations that implicitly include the influence of these factors in their records: for example, a station on the windward or eastern side of Lake Erie will have higher precipitation amounts than a station on the western side. These two different approaches, RCMs and ESDMs, are discussed in more detail in Chapters 4 and 5. The remainder of this chapter focuses on projections of regional climate change, providing an overview for each major world region, a discussion of the unique features of that region that moderate the influence of global change, and a summary of the broad-scale climate projections available for that region.

3.2 Climate Projections by Region

Although climate change is a global issue, its impacts vary widely by region. To resolve these important differences, even large international, global-scale assessments discuss projected changes and impacts by region. The first assessment to consider regional changes was the IPCC Second Assessment Report (SAR) (Watson et al. 1996). It divided the world up by climatological or ecological regions such as tundra, boreal forests, temperate forests, savannas, tropical forests, and grasslands. Later IPCC assessments use a more human-centric geographic and geopolitical definition of regions: Africa, Asia, Australasia, Europe, etc. (Solomon et al. 2007; Barros et al. 2014). Annex 1 of IPCC-AR5 (van Oldenborgh et al. 2013) provides detailed summaries of projected climate changes for each of these large-scale geographic regions, based on CMIP5 simulations (see Chapter 2) and IPCC-AR6 is expected to provide similar summaries based on CMIP6 simulations. To a large extent, this increased focus on regional projections has been driven by the increasing realization that regional-scale climate change is far more relevant to and significant for potential impacts (Urwin and Jordan 2008) and adaptation strategies (Adger et al. 2007) than global change as a whole. In turn, regional-scale assessments have spurred the development of downscaling techniques and analysis tools to generate high-resolution projections and translate them into impact-relevant information for stakeholders and decision makers.

This section describes the broad changes in mean climate that are consistent across multiple assessments and models for continental regions that have been studied in depth over the past few decades, and provides resources and references where users can find more information on each region. Unfortunately, assessments of regional-scale climate change projections and associated impacts on regions such as Africa, South Asia, Southeast Asia, and the Middle Eastern countries are relatively limited.

3.2.1 Projected Changes for North America

North America's climate conditions range from tropical to polar and from inland deserts to coastal rainforests. Due in part to the diversity of its climate regions, as well as to the relatively plentiful resources available for such analyses as compared to other regions, assessments and analyses of regional-scale climate projections for North America have been generated for individual towns and cities, states and provinces, and regions and countries. The state of California, for example, has performed a series of statewide climate assessments, the latest of which was completed in 2018 (Bedsworth et al. 2018). Other states, such as Colorado, have conducted statewide climate assessments (Lukas et al. 2014) while some cities, such as Chicago (Hayhoe et al. 2010), New York City (NYCPCC 2019) have done so as well. At the national scale, the US National Climate Assessment provides projections that include Canada and Mexico, while the Canada's Changing Climate Report (CCCR2019), (Bush and Lemmen 2019) includes projections and discussion of impacts for Canada specifically. The projections provided in the most recent of these reports, the Fourth National Climate Assessment (USGCRP 2017; 2018b) and CCCR2019 (Bush and Lemmen 2019), are based on CMIP5 simulations that had been previously downscaled using the LOCA ESDM for NCA4 and a combination of RCMs and an ESDM for CCCR2019.

Projected broad trends across North America include an increase in annual average temperature that ranges from 3.2 to 6.6 °C under the higher RCP8.5 scenario and 2.8 to 5.7 °C under the lower RCP4.5 scenario. Northern regions are projected to warm faster and to a greater extent; for all of Canada and Alaska, projected changes are expected to be twice those projected for the global mean. Changes in precipitation vary by both region and season, with consistent increases projected across more northern areas from the upper United States across Canada and Alaska, especially in winter and spring, while the southwest United States and Mexico are projected to experience significant drying during the same seasons. Soil moisture connects precipitation to temperature through evaporation; here, consistent and even stronger drying trends are projected across Mexico and the US Southwest and Rockies, especially in winter and spring, and across nearly the entire continent in summer (Figure 3.3).

A significant projected change for climate over North America is an increasing risk of extreme events. The NCA4 (Vose et al. 2017) projects that on average for the entire United States, the coldest and warmest day temperatures will increase by 2.8 °C and 5.5 °C respectively by the end of the century under the higher RCP8.5 scenario. By the end of the century the frequency of an extreme heat event, currently one in twenty years, is expected to be a yearly event. Analysis of RCM simulations for North America under the higher RCP8.5 scenario by end of the century project that five-day heat events will occur five to ten times more frequently over most of the contiguous United States and the number of 35 °C days will increase by between one to two months per year. Similarly, extremes in precipitation are expected to increase (Easterling et al. 2017).

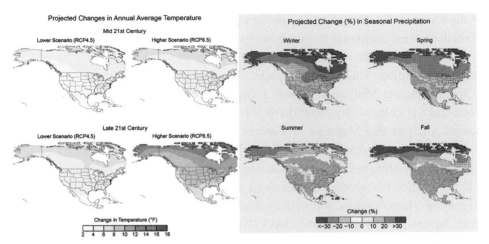

Figure 3.3 Projected changes in (left) average annual temperature (left), seasonal precipitation (middle) and seasonal soil moisture (right) for North America as simulated for a higher and lower future scenario (left) and a higher scenario (middle and right) by CMIP5 GCMs.
Source: Fourth US National Climate Assessment (USGCRP 2017)

Under the higher RCP8.5 scenario, extreme precipitation events with a return period of once every five years or over are expected to increase by factors of two to three over the entire country. The increase in extreme precipitation is projected to be accompanied by an increase in the number of dry days between the events (Zobel et al. 2018).

Unique to North America is the occurrence of large-scale forest fires in the western United States, including Alaska and Canada. The area burned by wildfires across the western United States has effectively doubled since the 1980s (reference the Southwest chapter of NCA4 Volume II) and is projected to continue to increase in the future (Wehner et al. 2017).

3.2.2 *Projected Changes for Central and South America*

Central and South America contains the world's largest tropical rainforest, vast mountainous regions, and large deserts. IPCC-AR5 provides a summary of the projected changes over South America (Magrin et al. 2014). Regional-scale projections are based on CMIP5 and have not been downscaled for the report. Brazil conducted a national climate assessment that was released in 2013 (PBMC 2013). One primary outcome of this assessment is that the temperature and precipitation changes over Amazonia could be between 5 °C and 6 °C by the end of the century compared to the end of the twentieth century and precipitation decreases of 40–50 percent are projected for this region. However, a concern noted in the report is the current rate of deforestation may be of higher concern for the immediate term compared to possible climate impacts in the future on this most extensive carbon sink on earth. Reports from three working groups ("Scientific Basis of Climate Change," "Impacts, Vulnerabilities and Adaptation," and "Climate Change Mitigation") were included in the report (PBMC 2013). The Food and Agricultural Organization (FAO) of the United Nations, under a program named Analysis and Mapping of Impacts under Climate Change for Adaptation and Food Security (AMICAF), performed an assessment of vulnerability of the agricultural systems in Peru (www.fao.org/in-action/amicaf/countries/per/en/). The UK Met Office has produced a series of country reports in 2011 and reports for Argentina (http://eprints.nottingham.ac.uk/2040/5/Argentina.pdf) and Mexico (http://eprints.nottingham.ac.uk/2040/17/Mexico.pdf). These assessments were primarily based on CMIP3 GCMs and IPCC-AR4.

Projected broad trends across South America show that, on average, the entire continent is expected to become warmer by the end of the century. Southeastern Amazonia is projected to warm more than the rest of the continent; increases of about 4.5 °C by the end of this century are projected for the higher RCP8.5 scenario (Figure 3.4) compared to 2 °C under the lower RCP4.5 scenario.

In terms of high-resolution projections, one set of RCM simulations using the lower RCP4.5 scenario driven by HadGEM2-ES model output (Imbach et al. 2018) shows trends similar to CMIP5 for mid-century, and a second study using an ESDM (Palomino-Lemus et al. 2018) projects for the higher RCP8.5 scenario a precipitation decrease of more than 50% compared with present climate conditions during the summer months across the middle of the continent and up the western side (Figure 3.5). The AMICAF assessment performed by the FAO for Peru using a statistical downscaling methodology, named MOSAICC (MOdelling System for

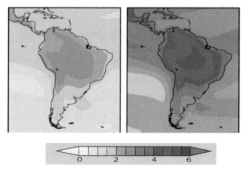

Figure 3.4 Annual average temperature range projected for the mid (left) and the end of the century (right) for South and Central for RCP8.5.
Source: IPCC-AR5 (Magrin et al. 2013)

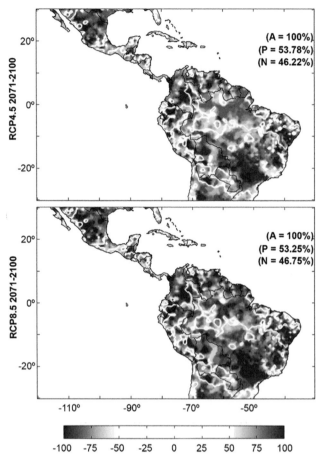

Figure 3.5 Projected changes in (top) average annual temperature and (bottom) southern hemisphere winter, June–August (2071–100) for Central and South America as simulated for a higher future scenario by twenty statistically down-scaled GCMs that participated in CMIP5. The numbers in the box represent the percentage of the areas with positive (P), negative (N), and total (A) change in precipitation.

The numbers' sources: IPCC AR5 (Magrin et al. 2014) and Palomino-Lemus et al. (2018)

Agricultural Impacts of Climate Change), projects the country as a whole to see an increase in temperature of between 2 and 3 °C to 4 to 6 °C by mid-century compared to 1971–2000.

In terms of unique features affecting South American climate, changes in the tree cover over the Amazon rainforest have been shown to have significant effect on regional and global precipitation (Avissar and Werth 2005). Frequency of extreme drought in South America has been increasing over the last few decades. From 2000 to 2015. three extreme droughts were recorded in Amazon (Erfanian et al. 2017). The most recent of these events, during the period 2015–16, was marked by extreme anomalies in precipitation in Amazonia during the southern hemisphere spring and winter. While a natural mode of anomalously warm sea-surface temperature over the Atlantic explained some of this rainfall deficit, modeling studies performed suggest a substantial role of increased greenhouse gases in the atmosphere as another driving cause (Erfanian et al. 2017). This type of enhancement of natural modes with climate due to warming (Dai 2012) is expected to increase in the future.

3.2.3 *Projected Changes for Europe*

Europe's climate conditions are primarily northern, but even still they range from polar to Mediterranean, with significant mountainous areas. As with North America, due to the relatively plentiful resources available for regional climate modeling and assessments compared to other regions, a number of regional-scale climate projections and assessments have been generated for individual cities, regions, and countries across Europe. At the continental scale, a comprehensive dynamic downscaling effort known as EURO-CORDEX (Coordinated Downscaling Experiment - European Domain) (Jacob et al. 2014; Kotlarski et al. 2014) provided the basis for a regional-scale climate assessment for the EU countries. Many country-specific assessments have also been performed within the EU. France, Germany and the UK in particular have performed extensive assessments (Havard et al. 2015; Brasseur et al. 2017; UKCP18 2018).

Future projections consistently indicate that, similar to the northern half of North America, Europe is expected to experience an increase in annual average temperature greater than what is projected for the global mean (Füssel et al. 2017). Figure 3.6 shows the projected changes in annual mean temperature and annual precipitation for the middle (2040) and end (2080) of the century for the higher RCP8.5 and very low RCP2.6 scenarios. Temperature and precipitation in summer over northern Europe are expected to increase under both scenarios, with higher scenarios leading to greater changes over the latter part of the century. Southern Europe, on the other hand, is expected to become both warmer *and* drier under both scenarios; again, projected changes are much larger by the end of the century and under the higher scenario. For southern Germany under the higher RCP8.5 scenario, for example, by the end of the century there could be more than thirty days with maximum temperature over 30 °C and as many as seven heat waves per year.

The Mediterranean region in particular is expected to experience widespread changes from the current favorable climate conditions. Projected changes include

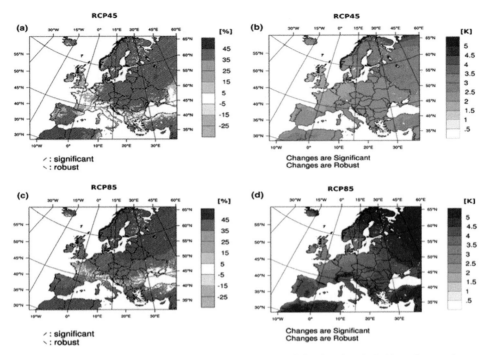

Figure 3.6 Projected changes in total annual precipitation in % (left) and annual average temperature (right) for Western Europe for a higher and lower future scenario (left) and a higher scenario (top and bottom) by RCM models used for Euro-CORDEX.
Source: Jacob et al. (2014)

higher temperatures during summer and decreasing precipitation. This is expected to lead to droughts and decrease in crop yields. Overall, this region is projected to be the most affected by changing climate in terms of the number of sectors impacted and the negative economic impact as a result. This region is also expected to experience the largest effect in Europe from climate changes happening across the borders affecting migration (Füssel et al. 2017).

3.2.4 Projected Changes for East Asia

East Asia, which includes the countries of China, Japan, South and North Korea, and Mongolia, encompasses a broad range of central deserts, the tallest mountain range in the world, and subtropical conditions in its southern coastal areas. Japan conducted assessments in 2014 and 2018 (Ministry of Environment, Japan 2018). China performed climate assessments in 2007, 2011, and 2015 (Wang et al. 2019b). These assessments were primarily based on the IPCC reports and use the same datasets (i.e. CMIP3, CMIP5) for generating the country-scale assessments.

The IPCC AR5 also provides a summary of the projected changes over East Asia (IPCC-AR5, Chapter 24, Hijioko et al. 2013).

In terms of regional features that are unique to East Asia, the Himalayas are a significant source of water for parts of East Asia and South Asia and play a critical role in the generation of monsoon wind patterns that produce most of the rainfall in South Asia The high mountain region also collects snow during the winter months from the winter westerly disturbances (Smith and Bookhagen 2018) and the glaciers in these high-altitude mountains provide the headwater for Yangtze and Mekong, the largest rivers in East Asia. The western part of East Asia is strongly affected by the East Asian monsoon, which brings much of the precipitation to this area during the warm summer monsoon and winter monsoon periods. Thus, projecting changes to the East Asian monsoon is of great significance to large population centers in China, Japan, and Korea.

For East Asia, projections consistently show the annual mean temperatures increasing by approximately 3–7 °C under the higher RCP8.5 scenario and 1–4 °C under the lower RCP4.5 scenario by the end of the century (Hong et al. 2017). Over Japan, under the higher RCP8.5 scenario, an increase in temperature is expected between 3.4 °C and 5.4 °C by the end of the century. Japan has experienced an increase in temperature of 1.19 °C over the past century, which is higher than that of the global average temperature rise during that period (Ministry of Environment 2018). The number of days with temperature over 35 °C for Japan is expected to increase and in particular an increase of fifty-four days is projected for the Okinawa region of Japan. China has similarly experienced a rapid increase in average temperature, with changes of 1.35 °C over the past century (Pioa et al. 2010). Using RCM simulations, Kang and Eltahir (2018) have estimated that by the end of the century the North China Plain region will experience increase extreme heat events and in particular the wet-bulb temperatures will exceed 30 °C for most of the region during summer, and 35 °C frequently.

Changes in seasonal precipitation based on CMIP5 projections under the higher RCP8.5 scenario downscaled using an RCM (Oh et al. 2014) project an increase in summer precipitation over eastern China, Korea, and Japan around 15 percent and nearly 50 percent over Korea for the period 2005–50 compared to 1971–2005 (Figure 3.7). This results from an increase in variability of the East Asian winter monsoon and an increase in the strength of the East Asian summer monsoon (Seo et al. 2013). Extreme events such as typhoons have been shown to be increasing in strength and producing more intense rainfalls in recent decades (Wang et al. 2015b). On the other hand, the frequency of typhoons is projected to decrease based on high-resolution RCM calculations over this region (Murakami et al. 2011).

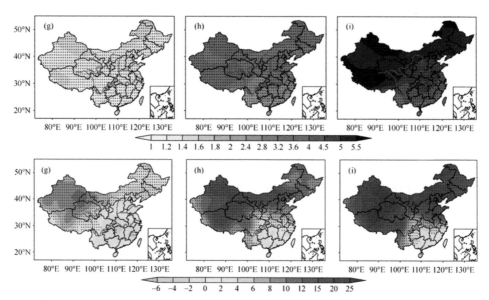

Figure 3.7 Multi-model ensemble average projections of temperature change (top panel) and precipitation (bottom panel) based on the CMIP5 simulations. Projections are shown for RCP8.5 for three time periods. The left panel presents the changes for the time period 2011–40, the middle panels for 2014–70 and the right panel for 2071–100 compared to a baseline of 1986–2005.
Source: Chong-Hai and Ying (2012).

Box 3.2
Impacts on Rice Cultivation in China

Rice is a staple in the Asian diet and has been cultivated in China for over ten thousand years. To feed a large size of population (approximately 1.4 billion in 2019), rice is widely cropped in China. The main rice-producing regions include the Northeast, Southeast, and Southwest. Climate change can influence the productivity of rice cultivation and thus affect food supply throughout this region. Researchers from the China Meteorological Administration, Chinese Academy of Meteorological Sciences, and other institutes used CMIP5 climate projections from twenty-three GCMs to understand how climate will change differently in northern and southern China and further investigate how will it affect local agriculture (Wang et al. 2014b). GCM projections were then spatially disaggregated.

Projections suggest precipitation will decrease in Guangxi and Guizhou, where rice is primarily grown in paddy fields, implying the potential for a significant negative impact on rice production in those regions. In Northeast China (Heilongjiang, Jilin, and Liaoning Provinces), radiation and temperature are projected to increase (Zhang

Continued

Box 3.2 (cont.)

et al. 2019), shortening the rice growing season by approximately three weeks but with a net positive impact on rice productivity.

Lv et al. (2018) investigated the impacts of climate change on regional rice production in China by combining simulations from 17 CMIP5 GCMs with a stochastic weather generator (the MarkSim method). This study agreed with these results, concluding that climate change will have a positive impact on rice crops in the Northeast China Plain but a net negative effect in central China. They also found significant shifts in the timing of annual rice planting will be needed to adapt to future climate.

3.2.5 Projected Changes for South Asia

South Asia, including India, Bangladesh, Pakistan, Nepal, Bhutan, Sri Lanka, and Maldives, is characterized primarily by a tropical and subtropical climate. Its precipitation is dominated by the southwest summer monsoon, which brings rain to most of the subcontinent, followed by the southeast winter monsoon, which provides rain to the southeast. A second important factor is the role of the Himalayas and the climate along the mountains, which sustains many of the largest rivers, such as the Ganges, Indus, and Brahmaputra, in the region and makes much of the land arable. Changes in snow- and ice-cover in the Himalayas and changes to the monsoon are driving factors for regional climate change and must be incorporated into assessing the impacts of climate change over this region. The only country in this region that conducted a climate assessment is India, in 2010 (INCCA, 2010). The UK Met Office has produced a series of country reports in 2011 and includes a report on India (www.metoffice.gov.uk/binaries/content/assets/metofficegovuk/pdf/research/climate-science/climate-observations-projections-and-impacts/india.pdf). This assessment, as the assessments done for Argentina and Mexico mentioned earlier, are primarily based on CMIP3 simulations and IPCC-AR4.

Bangladesh has already experienced an increase in daily average temperature of 0.47 °C during the period 1988–2017 (Md Hafijur Rahaman Khan et al. 2019). Using results from eleven different RCMs, the temperature over Bangladesh was projected to increase by over 4 °C by the end of the century under the higher RCP8.5 scenario (Fahad et al. 2018). The warming trend for India of 0.2 °C per decade has been observed over the period 1969–2005 (Basha et al. 2017). Average temperatures by end of the century are expected to increase by 5 °C for the higher RCP8.5 scenario when compared to historical averages from 1901 to 1960 (Basha et al. 2017), based on CMIP5 ensemble outputs (Figure 3.8).

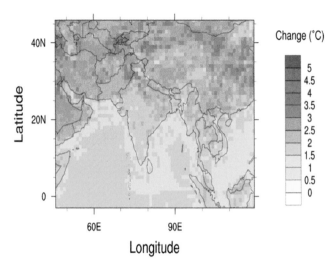

Figure 3.8 Projected change in surface temperature over South Asia by 2050 under the RCP8.5 scenario by mid-century. Results are from the CMIP5 ensemble and compared to a baseline temperature from 1986 to 2005.
Source: World Bank Climate Knowledge Portal (https://climateknowledgeportal.worldbank.org/country/india/climate-sector-health)

Both summer and winter precipitation are projected to increase by approximately 10 percent. There are, however, large variations within the CMIP5 ensemble regarding exactly how the southwest (summer) and southeast (winter) monsoons will change in the future over this region (Annamalai et al. 2007). For example, the summer monsoon is expected to intensify by the end of the century under the higher RCP8.5 scenario (Li et al. 2017). As shown in Figure 3.9, precipitation over India during all season is projected to increase by about 10% by mid-century under the lower RCP4.5 scenario and by more than 10 percent in spring, summer, and autumn for the higher RCP8.5 scenario, while winter precipitation is projected to decrease by more than 30 percent (Oh et al. 2014). Projected summer season precipitation increases are primarily concentrated in the northeastern sections of Bay of Bengal, while projected winter precipitation decreases occur over most of India.

Extreme heat over South Asia measured in terms of wet-bulb temperature is projected to increase to levels over 35 °C for regions in the Ganges river valley in North India, parts of Bangladesh and Sri Lanka, and the Indus valley region of Pakistan by the end of the century under the RCP8.5 scenario (Im et al. 2017). Extreme precipitation over India has shown an increasing trend over most of India and Southern India in particular from 1975 to 2015 (Mukherjee et al. 2018) and occurs primarily during the monsoon season. Mukherjee et al. (2018) define extreme precipitation as the event with highest observed value of annual mean

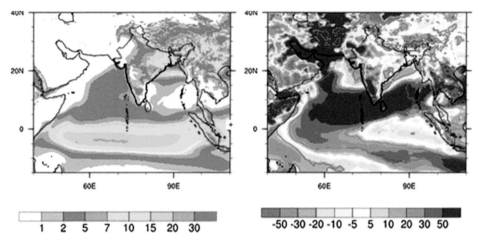

Figure 3.9 Projected changes in precipitation across South Asia for the RCP8.5 scenario and at the end of the century for the monsoon season (June, July, August, and September). Precipitation is shown in mm/day (left panel and label) and the changes are shown in percentage change (right panel and label). Projection generated using a 40 km Atmospheric general circulation model and CMIP5. Source: Woo et al. (2019)

precipitation event during 1975–2000. Using this threshold as the definition of extreme, they calculate Southern India will experience twelve such events by the end of the century under the higher RCP8.5 scenario and for all of India this number is projected to be four. These results were generated using the ensemble mean of the CMIP5 projections.

A unique aspect of hydrology over this region is the dependence on glacier melt water from high mountain regions for maintaining the flow in all the important rivers that provide irrigation for agriculture and support over a billion people. As a result, changes in snow fall or declining glaciers in the Himalayan mountains and the high mountain regions that extend from Hindu Kush to Himalayas are critical. A recent study that utilized data from spy satellites in addition to data sets used in previous studies has concluded that the glacier melting rate in this region has accelerated and the loss rates are double in 2000-2016 compared to 1975–2000 (Maurer et al. 2019).

3.2.6 Projected Changes for Africa

Climate across African continent varies from the Sahara and sub-Saharan regions to tropical rainforests. As such, the continent is expected to experience varying degrees of effects from climate change. Most of the continent lies within the tropics and the climate is influenced by its position on either side of the equator.

The intertropical convergence zone (ITCZ) that is positioned along the middle of the continent strongly affects its climate. The regions in the ITCZ itself also experience significant rainfall and have a wet tropical climate, whereas the regions north and south of this zone have semiarid to desert conditions.

South Africa has conducted two national-scale climate assessments – the 2nd National Climate assessment was published in 2016 (Department of Environmental Affairs 2017). By the end of the century under the higher RCP8.5 scenario, the Sahel region is projected to have temperature increases of approximately 6 °C (IPCC-AR5, chapter 22; Niang et al. 2014) based on downscaling calculations performed at a spatial resolution of 50 km (CORDEX-AFRICA). Over Central Africa, the projected average temperature change by the end of the century is 2.3 °C for the lower RCP4.5 and 4.1 °C for the higher RCP8.5 scenario (Aloysius et al. 2016). Another study on the projected temperature changes calculated using dynamic downscaling with a different model (COSMO-CLM) is in general agreement with the ESM projections. However, RCM results produce a +1 °C higher warming for South Africa during the Austral summer months and Sahel region during the Austral winter (Dosio and Panitz 2016) than estimated with ESM. These projections are shown in Figure 3.10.

Precipitation over many parts of Africa is influenced by monsoons on both sides of the continent. CMIP5 simulations project less rainfall along the western

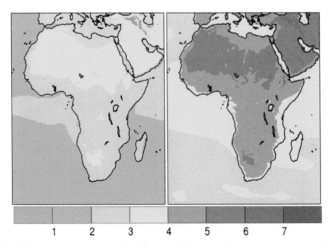

Figure 3.10 Ensemble mean projections of temperature change over Africa produced using an RCM and nine CMIP5 GCMs. The label shows the temperature change in degrees C compared to a baseline year of 1971–2000 by the end of the century. The left-hand side figure is for the lower RCP4.5 and the right-hand figure is for the higher RCP8.5 scenario.

Source: Swedish Meteorological and Hydrological Institute (SMHI), Global Warming Levels: www.smhi.se/en/climate/global-warming-levels/africa/2_0C/year/temperature

Sahel region and increased rainfall along the central and eastern Sahel regions by mid-century under the lower RCP4.5 scenario (Monerie et al. 2012). An increase in precipitation of more than 30 percent is projected for the central and eastern Sahel by the end of the century under the higher RCP8.5 scenario, while the western Sahel is projected to experience a small decrease. Using the CMIP5 ensemble under the higher RCP8.5 scenario, Central Africa is projected to experience an increase in precipitation on the order of 10–20 percent by the end of the century during October–March and a very small change during the April–September period. Under the higher RCP8.5 scenario, Southern Africa is projected to experience a decrease in rainfall of up to 20 percent during the April–September period and smaller decreases during the October–March period (Figure 3.11).

A distinct issue for Africa is the potential for increases in the frequency and extent of droughts. Analysis of drought frequency and duration based on IPCC-AR4 show the potential for increase in short term drought frequency across West Africa, South Africa, and Central Africa, with West Africa being most impacted. Long-term droughts are also projected to increase in West and South Africa and remain about the same for Central Africa (Sheffield and Wood 2007). A CMIP5-based analysis by Lu et al. (2019) shows an increase in both short- and long-term droughts for South, Central, and North Africa for both the lower RCP4.5 and higher RCP8.5 scenarios by the end of the century compared to a baseline of 1976–2005. This analysis projects the biggest changes in the frequency of droughts for South Africa compared to other parts of Africa.

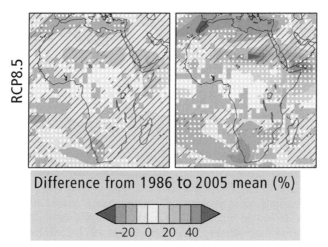

Figure 3.11 CMIP5 ensemble projections of precipitation changes in Africa by mid and end of the century under the higher RCP 8.5 scenario.
Source: IPCC-AR5, Chapter 22 (Niang et al. 2014)

Box 3.3
Ethiopia Future Water Availability Could Spur
Agriculture and Hydropower

The Highlands of the Ethiopian Blue Nile Basin (BNB) are a major source of Nile River discharge. The highly erosive lands contribute significant sediment loads to the Nile's waters, which challenges downstream reservoir operations by decreasing capacity over time. Researchers from Virginia Tech and University of Maryland, USA, and Abay Basin Authority and International Water Management Institute, Ethiopia, used CMIP5 simulations to understand how the basin's climate will likely change in the future and how these changes will impact water resources downstream. Using four RCP scenarios, they applied bias correction–spatial disaggregation methods via the QMAP R package (Gudmundsson et al. 2012) to downscale future climate projections.

The results indicate the region will experience higher precipitation and temperatures in the future, which translate to a longer growing season, increased annual streamflow, and increased sediment concentrations during early and late monsoon season. The largest streamflow and sediment increases in summer to early fall result from increased frontal-type orographic storms as well as increased precipitation early in the monsoon period.

For the region, the future climate could significantly improve agricultural capacity through the extended growing season with adequate rainwater for crops. However, the increases in sediment loads will require reservoir operators to implement strategies such as soil conservation measures or strategic bypassing of flows during early season periods when sediment concentrations are particularly high.

Source: https://phys.org/news/2016-10-climate-ethiopia-country-access.html

---

3.2.7 Projected Changes for Australia

The climate of Australia is dominated by the desert and semiarid regions over most of the continent, with wetter and tropical conditions in northern locations along the coast bordering the Indian Ocean. The most recent assessment of climate change over Australia was performed by the CSIRO (Commonwealth Scientific and Industrial Research Organization) and the Bureau of Meteorology (BOM) (CSIRO and Bureau of Meteorology 2015).

Projections for Australia show an average temperature increase of 1.4–2.7 °C by end of century under the lower RCP4.5 scenario and between 2.8 °C and 5.1°C under the higher RCP8.5 scenario. Figure 3.12 shows that the changes over Northern Australia will be much higher than Southern Australia. Precipitation over most of the Australian continent is projected to decrease, with reductions of approximately 50 percent projected for Southwest Australia

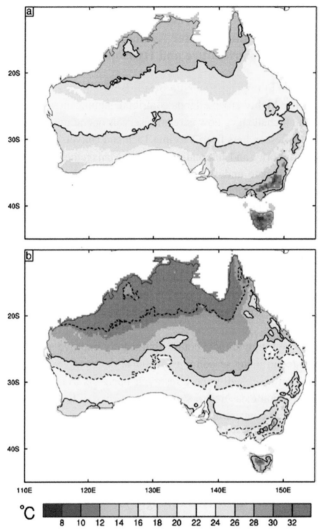

Figure 3.12 Annual mean temperature under (a) current climate and (b) for the end of the century for Australia. The projected temperatures are estimated by adding the median of CMIP5 model projections to current temperatures.
Source: CSIRO and Bureau of Meteorology (2015)

for the winter months (CSIRO and Bureau of Meteorology 2015). Region-wide averages mask the large variability over the continent, with southwestern Australia experiencing a decrease of approximately 15 percent while the rest of the continent is projected to see little to no change by the end of the century under the higher RCP8.5 scenario (Figure 3.13)

Australia is experiencing increasing brush fires and the number of days with fire has increased from twenty-four to twenty-eight sites over Australia from

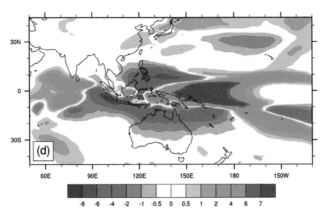

Figure 3.13 Projected changes in summer rainfall over Australia and adjacent regions for the period between 2066–100 compared to 1971–2005.The results were obtained using GCM developed by the CSIRO.
Source: Suppiah et al. (2013)

1973–2010. It is projected that many parts of Southern and Eastern Australia where the fuel availability is higher, will experience more incidents in the future. Extreme temperatures are expected to rise around the country. Major cities such as Perth and Adelaide are projected to see an increase of 50 percent in the number of days with temperature over 40 °C and 35 °C respectively.

3.3 Regional Projections of Sea Level Change and Marine Temperature

For many coastal communities, projected changes in sea level may be the most important and relevant climate impact. As discussed in Chapter 1, over 700 million people currently live in low elevation coastal zones. Coastal areas are also home to two-thirds of the world's largest cities and untold trillions of dollars of valuable infrastructure, from ports and naval bases to oil and gas extracting and refining facilities.

At the global scale, sea level has already risen by about 16–21 cm and that rate of rise is accelerating (Figure 3.14). Over the coming century, global sea level is very likely to rise another 0.3–1.3 m, and for higher scenarios as much as 2.4 m is physically possible (Hayhoe et al. 2017). Future projections of sea level rise due to both thermal expansion and the melting of land-based ice are commonly generated by empirical models. At the regional scale, however, global sea level rise can be significantly modified by uplift or subsidence.

As shown in Figure 3.14, observed trends in regional sea level range from a rise of more than 9 mm per year in the US Gulf Coast, where withdrawal of oil, gas,

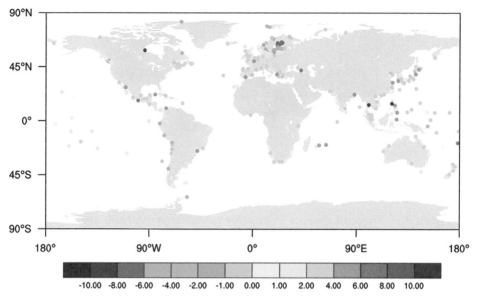

Figure 3.14 Observed trends in local sea level rise (mm/year) show that global change can be significantly moderated by coastal subsidence (e.g., the Gulf Coast of North America, eastern Asia) or uplift (e.g., the south coast of Alaska, northern Scandinavia, and the western coast of South America).
Source: NOAA Tides & Currents www.tidesandcurrents.noaa.gov/sltrends/

and water from underground reservoirs is driving large-scale subsidence, to a decrease of more than 9 mm per year along the southern coast of Alaska, where the land is still rebounding from the melting of the ice sheets during the last glacial maximum.

While there are a few examples of using a regional-scale ocean model to downscale sea level rise projections (Li et al. 2014) from global-scale models, similar to the concept of dynamic downscaling, much more effort has been invested in statistical bias-correcting of sea level rise (Cui et al. 1995). IPCC (Nicholls et al. 2014) recommends that regional sea level change be calculated as the sum of the following components:

$$\Delta RSL = \Delta SL_G + \Delta SL_{RM} + \Delta SL_{RG} + \Delta SL_{RLM} \tag{1}$$

where ΔRSL is the total change in relative sea level at a given location, ΔSL_G is the change in global mean sea level, ΔSL_{RM} is the regional variation of sea level rise due to regional-scale ocean circulation features, warming, and other features that are defined by regional ocean circulation and properties, ΔSL_{RG} is the part of sea level change that can be attributed to regional variations in the gravitational field of the earth, and ΔSL_{RLM} is due to any changes in the topography from example due to earthquakes. Each of these terms can be estimated via various statistical downscaling approaches and as detailed in Nicholls et al. (2014). Simulation models that include all these changes are also available, (SlimCLIM, Warrick, 2009). More simplified approaches are available as the digital coast mapper available from NOAA for the United States (https://coast.noaa.gov/digitalcoast) and for various parts of the globe (http://globalfloodmap.org).

4

Dynamical Downscaling

Dynamical downscaling uses high-resolution regional climate models (RCMs) to bias-correct and downscale GCM output. This chapter discusses the models and methods used in dynamical downscaling. It provides an overview of the basic physics used in RCMs, and how this is similar to and differs from that used in global models. It also discusses the methods and metrics used to evaluate RCMs, and how projections from RCMs can be used to assess climate impacts at the regional scale.

4.1 Regional vs. Global Climate Models

Regional climate models (RCMs) are physical models of the climate system that cover a limited geographic area. The earliest RCMs were initially based on weather-prediction models, but RCMs today have many new components that have been added to improve their ability to simulate the atmosphere and land surface over longer timescales, from years to decades.

Although RCMs cover a limited area at relatively high resolution, they are still very similar to GCMs in many ways. Both are physically based models that can directly simulate the many different processes affecting the atmosphere, ocean, and land surface at the spatial scale of the grid cells being used. As discussed in Chapter 2, both types of models use a series of equations and parameters to describe smaller-scale processes, such as precipitation, that the model cannot resolve. Both GCMs and RCMs produce three-dimensional output that includes surface temperature and precipitation as well as vertical temperature, humidity, winds, cloudiness, and more. Like GCMs, RCMs continue to evolve, both by increasing their resolution (as fine as 1–2 km in some experimental RCMs and typically around 10–12 km for RCMs used to generate long-term climate projections) as well as by improving the representation of physical processes in the model.

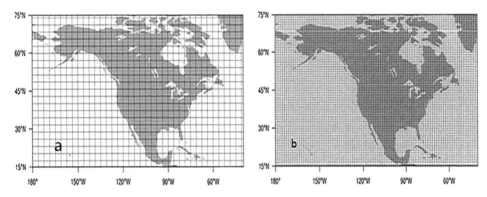

Figure 4.1 Regional-scale climate models (RCMs) provide higher spatial resolution than GCM grid cells: (a) shows the typical size and number of horizontal grid cells in a GCM for North America (~100 km grid cell size); (b) shows the same region with approximately 100 times more grid cells. The grid cell size in (b) is approximately 10 km (for clarity, only every third grid boundary is drawn).

One of the key differences between GCMs and RCMs is the fact that global models are continuous, covering the entire globe, so they do not have horizontal edges. In contrast, regional models cover a limited area, usually a rectangle centered over a specific region (Figure 4.1). This means that the edges of the regional model must be regularly updated with output from a global model. This can be accomplished in two ways. The first approach uses pre-calculated GCM output to update the boundary conditions on what is known as a limited-area model (LAM) or RCM every three to twenty-four hours, depending on the model (Figure 4.1). The second approach interactively embeds an RCM within a GCM by using a variable-resolution model that provides coarse resolution at the global scale and "zooms in" to provide higher resolution in the area of interest (Figure 4.2).

Variable-resolution models that combine global-scale climate models with embedded regional grids have been part of GCM development for nearly two decades, but they are less commonly used than stand-alone RCMs (Fox-Rabinovitz et al. 2005; Mearns et al. 2014; Rhoades et al. 2016). Figure 4.2 illustrates such a model, where a global-scale grid of 90 km zooms in over the contiguous United States (CONUS) to a much finer grid size of 9 km. The major advantage of variable-resolution models is that they allow for a strong, real-time coupling between atmosphere, ocean, and land processes from regional to global scales. The main reason they are not more commonly used, however, is that variable-resolution GCMs are extremely computationally expensive, requiring much more time to run as compared to a stand-alone RCM using preexisting GCM output. For that reason, output from variable-resolution GCMs is generally neither available nor has simulations available corresponding to future scenarios

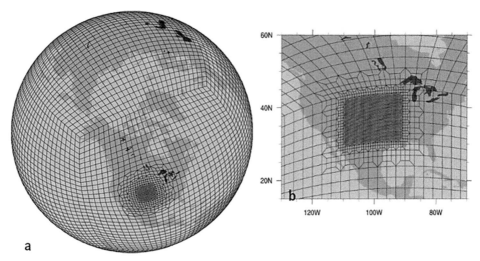

Figure 4.2 Regional refinement in a global model with higher density of grid cells and hence better resolution of physical processes over CONUS. The global model shown in (a) has approximate grid size of 90 km and the regionally refined grid over CONUS (b) has a resolution of 9 km.

that are suitable for the type of applications discussed in the rest of the book. In the future, however, as computational power increases, these variable-resolution models may become more commonly used. For consistency, this book uses "RCM" to refer exclusively to the former approach; but both approaches are described as "dynamical downscaling" because they directly simulate the dynamics of the regional climate system.

The most important difference between GCMs and RCMs, however, is their spatial resolution. As shown in Figure 4.1, dynamical downscaling using RCMs can achieve spatial resolutions that are ten times higher than GCMs over smaller regional areas. Higher spatial resolution means that RCMs are better able to model complex terrain, including coastlines, as well as to better represent processes that are affected by changes in land use and land cover; often, they can better simulate hydrology and other processes at scales of interest to decision makers as well.

The higher spatial resolution of RCMs means they are better able to resolve and represent some key physical processes. Some sub-grid processes at the spatial resolution of an GCM can be explicitly resolved at the spatial resolution of an RCM. Thus, RCMs and GCMs differ in what could be considered sub-grid, as they are designed to model atmospheric processes at different spatial scales. For example, RCMs use parameterizations to represent the mixing process within the atmospheric boundary layer, which includes approximately the lowest two kilometers of the atmosphere. The atmospheric boundary layer is influenced by incoming solar radiation and by conditions on the surface of the Earth over the twenty-four-hour daily cycle. GCMs do not have a parameterization to represent mixing within the atmospheric boundary layer, as the grid resolution is much coarser than the timescales of relevance to boundary layer processes in the lower atmosphere. Similarly, the sub-grid-scale representations of clouds and, in general, any physical process that depends on the spatial resolution of the model, have different representations in GCMs and RCMs. However, in general, both GCMs and RCMs represent all the physics of the atmosphere using parameterizations that are appropriate for the spatial resolution of the model.

4.2 The Physics of Regional Climate Models

As mentioned, RCMs first evolved from weather forecast models about three decades ago. The first studies simply converted an existing weather forecast model, the Mesoscale Modeling version 4 (MM4), into a regional-scale climate model by using GCM output to set conditions at the edges of the model instead of observations used to initialize weather forecasts (Dickinson et al. 1989; Giorgi and Bates 1989). Later, the model was modified to improve its ability to simulate the behavior of the atmosphere over timescales longer than those typically relevant to weather forecasting.

The climate version of MM4, developed by Giorgi and colleagues, became known as RegCM2. Since its early days, it has undergone extensive development (Giorgi et al. 1993a, 1993b; Pal et al. 2007), which has resulted in several new versions (RegCM3, 4, and 5). Various versions of the RegCM model and the Weather Research Forecast (WRF) model, a widely used limited-area model for both mesoscale weather and climate analyses developed at the National Center for Atmospheric Research (NCAR), are still used by the dynamical downscaling community for producing high-resolution, numerical-model-based projections of future climate (Liang et al. 2012; Mearns et al. 2013; Wang and Kotamarthi 2015; Zobel et al. 2018). Other RCMs in the rapidly expanding set of models used for dynamical downscaling include the Hadley Centre model HadRM3P (Li et al. 2015), the REMO-RCM developed by the Max-Planck Institute, and the community model from Germany, COSMO-CLM (Rockel et al. 2008).

What types of physical processes are included in an RCM? Figure 4.3 illustrates the most common of these. First, the models calculate important aspects of atmospheric circulation including high and low pressure systems, wind, convection, the formation and dissipation of clouds and storms, precipitation, and other atmospheric phenomena we associate with weather and climate. Second, they capture variables and processes that couple the land to the atmosphere. These include soil moisture, heating and cooling of soil layers, the exchange of water

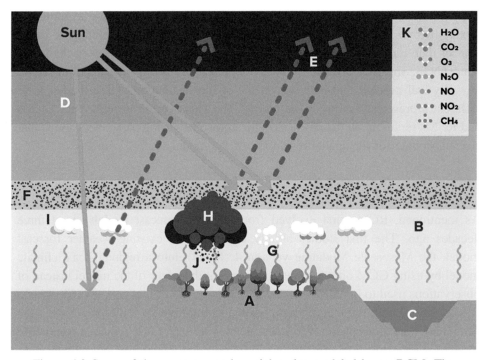

Figure 4.3 Some of the processes and spatial scales modeled by an RCM. The models calculate circulation in the atmosphere, cloud processes, precipitation, and land–atmospheric and ocean–atmospheric processes on a limited portion of the Earth: (A) represents biological processes that affect the exchange of water vapor and atmospheric gases between the soil and atmosphere and between plants and the atmosphere; process (B) represents the mechanical and heat-driven mixing of the atmosphere from the surface to higher altitudes; (C) represents the surface runoff of water driven by gravity; (D) represents incoming short-wavelength solar radiation; (E) represents the long-wavelength radiation emitted into space by the Earth and atmospheric gases; (F) represents the formation and removal of aerosols in the atmosphere from precursor gases or directly produced by human activities, such as burning fossil fuels; (G) represents the microphysical properties of cloud formation, formation of cloud droplets, droplet growth and precipitation; (H) represents the formation of deep convective clouds in the atmosphere that result from vigorous mixing and movement of air rapidly from the lower atmosphere to the upper atmosphere; (I) represents the formation of shallow cumulus clouds; (J) represents the entrainment of air and moisture into clouds; and (K) represents the life cycle of various atmospheric gases that directly or indirectly affect the radiation balance of the earth.

between vegetation and atmosphere, and the flow of water on the surface and in the ground. The biosphere is typically represented using observational data that are fixed (although some simulations include a dynamic vegetation model that changes with climate over time) and the ocean surface is represented by surface temperature values obtained from observations for the historical period and ocean model simulations for future conditions. Finally, the transfer of energy, or radiative transfer, through the atmosphere is represented through reducing incoming energy from the sun and outgoing energy from the earth into a series of wavelength bands to allow for more efficient calculations.

How do RCMs actually model these processes? Similar to GCMs, they begin by solving a set of what is known as "governing equations." These equations represent atmospheric motion (i.e., advection), energy transfer, and the transport or movement of water vapor and other gases and particles through the atmosphere. They include what are known as the "primitive equations," which describe the conservation of momentum and the temperature and density of air. For more details on the physical equations that form the basis of a typical RCM, see Dutton (1995).

Modeling these equations is complicated by the fact that, as in GCMs, analytical methods can only integrate the equations in one dimension, with simplifying assumptions. To obtain a general solution, the models use a numerical solver to "discretize" the equation – i.e., convert a differential equation that is continuous in space and time into a discrete form or difference equation for which a numerical solution can be calculated on a computer. For example, as shown in Figures 4.1 and 4.2, the space over which the model needs to solve these equations is divided into a series of rectangular boxes. Numerical solvers provide solutions to the equations at each of the corners of these grid boxes. The boxes are referred to as "grid cells" and the distance between the edges of each individual box is the "grid size." The grid cells in this case (Figures 4.1 and 4.2) are rectangular and the grid sizes are represented as two dimensions along the lines of latitude and longitude. A more complex grid system is shown in Figure 4.2, which uses a hexagonal mesh for discretization of space. A detailed description of discretization processes and of additional equations that are solved in the WRF model, one of the most widely used models for regional-scale dynamical downscaling, can be found in Skamarock et al. (2008).

While most of the basic equations used in GCMs and RCMs are similar if not identical, one significant difference relates to the hydrostatic assumption. The hydrostatic assumption implies that the change in the atmospheric pressure with elevation is approximately balanced by gravity. This should be true if there are no large lateral atmospheric motions. In a GCM, which has a much coarser spatial resolution, this approximation will hold within a grid cell, as the horizontal scales of motion resolved are much greater than those that affect the hydrostatic assumption. This assumption is relaxed in RCMs as the grid size becomes smaller; the horizontal scales of motions resolved are on the order of a few kilometers, and the hydrostatic assumption is no longer valid.

As with GCMs, some physical processes within RCMs are not modeled directly, but rather parameterized: represented by a simplified approach that describes the behavior of the system in aggregate. Processes are typically parameterized if they cannot be resolved at the spatial scales of the model, or if they are used for the discretization scheme described, or if they are poorly understood. Important aspects of climate that are parameterized in RCMs include boundary- (near-surface-) layer physics, convective and cumulus clouds, and radiative transfer. Each of these parameterizations is a sub-model that represents the bulk features of the phenomenon and its impact on the rest of the atmosphere. These model parameters are primarily set based on observational studies conducted to understand the specific process or phenomenon in the real world. Since these parameters are obtained by fitting a model to observations of a given phenomenon, they have uncertainties associated with them. All of the processes identified in Figure 4.3 happen at spatial scales smaller than the typical grid size used in both RCMs and GCMs.

As with any physical model, RCMs are tested under limited sets of conditions based on the availability of observational and paleoclimate data. Because of this, there is uncertainty in using these models over a large range of conditions and many geographical and climatic conditions. These uncertainties are referred to as parametric uncertainty (when they deal with how the model parameterizes processes) or structural uncertainty (when they relate to what processes are or are not included in the model and whether they are represented accurately). These sources of uncertainty are discussed in more detail and methods for accounting for these uncertainties are presented in Chapter 7.

4.3 Outputs from Dynamical Downscaling Models

Dynamical downscaling outputs generally relevant to impact assessments include temperature, humidity, soil temperature and moisture, geopotential height, pressure, atmospheric water vapor, wind speeds and directions, precipitation (accumulation over some pre-selected time periods), type of precipitation (i.e., snow vs. rain), temperature at and above the surface of the Earth, incoming and outgoing solar radiation, cloud cover, sensible and latent heat flux, and many other variables that are required to compute the dynamic state of the atmosphere (Table 4.1). Typical model output consists of 60–100 variables. The atmospheric variables that vary with both height and horizontal location correspond to multiple layers in the atmosphere extending from about 2 m above the surface to the top of the troposphere (around 16 km), with a value at each of the grid cells. Subsurface profiles of soil moisture and soil temperature are available to a depth of 2 m, depending on the method used to represent the land process in the model.

Table 4.1 *Types of outputs available from RCMs and their use in impacts assessments and analysis.*

Variable	Use for Analysis	Spatial	Temporal
Temperature (at 2 m above the surface)	Calculating lowest layer (surface) temperature at a grid location.	Available at every grid location, values are at either the corner of the grid cell (lower left) or center of the grid cell.	Availability varies with model grid size. Models with smaller grid sizes archive data every hour to three hours (~10 km grid cells); models with larger grid cells (~50 km) archive data four times a day.
Atmospheric Temperature Profile	Estimating temperature changes in different layers of the atmosphere and changes to atmospheric stability.	Available at every grid location and each model altitude grid level. RCMs generally go to the top of the troposphere (~16 km). The number of grid levels varies from twenty-six to over 40, and usually have more grid levels in the lower atmosphere (below 2 km) than in the upper atmosphere.	Availability varies with model grid size. Models with smaller grid sizes archive data every one to three hours (~10 km grid cells); models with larger grid cells (~50 km) archive data four times a day.
Precipitation (liquid and solid; caused by large-scale and small-scale atmospheric phenomena)	Estimating precipitation rates and phases and their changes over time.	Available at every grid location as a grid cell average (mm/km^2).	Availability is generally as a daily average, models with higher spatial resolutions (10 km or less) archive precipitation data from every 1–3 hrs.
Snow (snow water equivalent, and snow depth)	Estimating winter snowpack water content, snow-covered areas, and their changes.	Available at every grid location that has as a grid average.	Availability is generally as a daily average, models with higher spatial resolutions (10 km or less) archive precipitation data from every 1–3 hrs.

Table 4.1 (*cont.*)

Variable	Use for Analysis	Spatial	Temporal
Soil Properties (soil moisture, soil temperature)	Estimating drought conditions, rainfall/ runoff partitioning, and wildland fire potential.	Soil moisture is provided as a grid average and soil temperature is either from the corner of the grid cell or mid-point of the model grid. Available at every model grid cell. different depths (up to 2 m below the ground) in the soil.	Availability varies with model grid size. Models with smaller grid sizes archive data every hour to three hours (~10 km grid cells) and models with larger grid cells (~50 km) archive data four times per day.
Wind (surface and vertical profiles; speed and direction; along the meridional (longitude), zonal (latitude), and vertical (altitude) direction)	Calculating wind speeds maxima, minima, and directions. Applications range from wind energy availability to extreme events.	Wind components (u, v, w) are available at every grid point of the model either at the corner of the grid cell or mid-point of the model grid from the surface of the model (2 m) to the top of the model domain.	Availability varies with model grid size. Models with smaller grid sizes archive data every one to three hours (~10 km grid cells) and models with larger grid cells (~50 km) archive data four times per day.
Cloud Properties (cloud fraction and liquid-water content)	Understanding relationship between cloud properties and changes in precipitation intensity, duration, and location.	Cloud fraction and liquid water content of the clouds are available at every grid point in the model. Cloud bottom height, and cloud top height can be calculated with the information.	Availability varies with model grid size. Models with smaller grid sizes archive data every hour to three hours (~10 km grid cells) and models with larger grid cells (~50 km) archive data four times per day.
Planetary Boundary Layer Properties (height, surface energy, and moisture fluxes)	Analysis of droughts, fires, and agricultural impacts and changes to atmospheric stability.	Values are for each grid cell and single values giving the height of the boundary layer, flux values near the surface.	Availability varies with model grid size. Models with smaller grid sizes archive data every hour to three hours (~10 km grid cells) and models with larger grid cells (~50 km) archive data four times per day.

Table 4.1 (*cont.*)

Variable	Use for Analysis	Spatial	Temporal
Atmospheric Water Vapor	Analysis of phenomena such as atmospheric rivers, mid- and low-level jets over the Midwest that bring moisture over the midwestern states and lead to precipitation events.	Water vapor content of the atmosphere are generally available at each grid cell in the model.	Availability is similar to the variables described.

Similar to GCMs, RCM model output is saved in compressed data formats specifically developed by the atmospheric modeling community over the past couple of decades. The most commonly used data format for storing output from both GCMs and RCMs is NetCDF, which stands for Network Common Data Form. NetCDF files are self-describing files, meaning that each file comes with its own metafile describing the names of variables, the dimension of the data, spatial-location information (e.g., latitude and longitude), and units; thus, no additional files are usually required for performing geospatial analysis. However, data stored in NetCDF are in binary format and hence not readily readable. It requires the use of custom-developed tools that contain NetCDF data reading and writing protocols. Most programming languages and computing environments, including Matlab, R, and ArcGIS, are able to read NetCDF files. Many other tools for reading and writing data and for visualization, such as NASA's Panoply program, were developed under open-source licensing agreements and are available free of cost. Some commercially licensed software used for assessing and visualizing geospatial datasets and analysis has built-in or third-party codes that make it feasible to import NetCDF data files.

4.4 Workflow for Performing Dynamically Downscaled Simulations

Running a dynamically downscaled simulation consists of three main steps: selecting the GCM simulation (including the historical and/or future scenario) that will provide the boundary conditions for downscaling; setting up the RCM; then running the RCM, and archiving its output. Each of these steps, or phases, is described in more detail below.

In Phase 1, the question of which GCM to use as input for a given simulation is often constrained by logistics: which GCMs have archived simulations that represent the historical climate and future projections for a selected set of scenarios at a temporal resolution of six hours or higher? The GCMs also must have archived sea-surface temperatures if the RCM model domain extends over surrounding ocean. Then, the GCM data have to be pre-processed so they can be used as input to the RCM. This is typically accomplished using software and procedures that are well established for each RCM. A frequently used step in preparing the input and boundary conditions for the RCM at this stage is to apply bias correction (Xu and Yang 2012; Bruyère et al. 2014; Wang and Kotamarthi 2015). The goal of applying a correction to the regional-scale climate model input fields from the GCM is to correct the errors in the mean fields of selected atmospheric variables while preserving the longer-term climate variability of the of the GCM input fields. The calculated bias between the mean fields of the historical climate and the GCM simulations is removed from the GCM projections for a future time period as well, and these corrected fields are then used to perform the downscaled simulations with the regional-scale model.

Phase 2 is setting up the RCM model: choosing which physics package to use, how to run the model, and the interval at which the model archives output. The choice of physics package is generally based on testing the model with a physics option, such as a convective cloud scheme, evaluating the model with observations, and deciding if the model produces reasonable results for precipitation for the study area with that choice of the convective cloud parameterization. Another option for running the model is to use nudging (Mabuchi et al. 2002; Miguez-Macho et al. 2005; Lo et al. 2008; Wang and Kotamarthi 2013). Nudging is a process by which selected large-scale atmospheric phenomena, such as temperature and moisture that are driven by large-scale phenomena in the GCM, are preserved and used to drive the smaller-scale motions that are resolved by the RCMs and to prevent the RCM simulations drifting too far from the input GCM conditions. For example, temperature and water vapor at 500 millibars and above in the atmosphere in the RCMs are nudged back to that from the GCM boundary conditions over the entire model domain every six hours or at the same frequency with which GCM fields are input into the RCM as boundary conditions. The advantage of performing nudging is that the RCM fields at the larger spatial scales remain close to the GCM fields while the downscaling or higher-resolution modeling works to resolve smaller-scale features that are associated with these large-scale fields. The disadvantage is that the downscaled fields have reduced variability in the output in the model calculated fields at the surface (Wang and Kotamarthi 2014).

RCMs produce an enormous amount of data; to keep storage requirements within reasonable limits, it is necessary to save only a small fraction of the data. For example, an RCM that covers the CONUS at a spatial resolution of 4 km with twenty-eight layers through the atmospheric column will have approximately 30 million grid cells. Model outputs for most variables are produced at each of these grid cells at a time step used to solve the equations, which is generally less than a minute for a model operating at these spatial resolutions. This corresponds to a total output of the order of several hundred terabytes per day of simulations. To keep the storage requirements reasonable, model output is typically saved only once every so often, as dictated by the problem being solved. For example, if only the daily maximum and minimum values are required, saving once a day is all that is necessary: the file size will be in the range of a few gigabytes per single day of simulation output and data size for multiple years of data will be in the range of a few hundred terabytes. For other applications that require sub-daily or hourly data, the amount of data stored will correspondingly increase and the frequency of data output will be more often limited by available storage.

Phase 3 consists simply of carrying out the computations and storing the output from the model: but this is often the longest step, as these models are very computationally demanding. For example, a simulation of most of the North American continent at 12-km horizontal grid cell size and twenty-eight vertical layers, with the top of the atmosphere set at 100 mbar, has approximately 8.5 million grid cells. On a 64-node computer in which each node contains a 16-core an IBM SP/Q chip rated at 1.6 gigaflops it requires approximately 0.6 million core hours of computing time to complete a 1-year simulation see (Figure 4.4).

These three steps are generally repeated with multiple combinations of GCMs and scenarios, and sometimes even multiple RCMs, to cover the range of the structural and parametric uncertainty in these models. For the purpose of downscaling, a general approach to capturing model uncertainty is to select a range of GCMs that cover this uncertainty space and use these to develop an ensemble of dynamically downscaled datasets. As discussed in more detail in Chapter 7, this could be done by selecting GCMs with different values of climate sensitivity (Zobel et al. 2017). Selecting a range of RCMs or different physics packages or choices of sub-grid scale parameterizations within the same RCM also helps to cover the range of uncertainties in the higher-resolution modeling. Finally, scenario uncertainty is generally addressed by using GCM simulations corresponding to a range of future scenarios, such as a lower RCP4.5 and a higher RCP8.5 scenario (Zobel et al. 2017). Sources of uncertainty in high-resolution climate projections and best practices in addressing them when using this information to quantify climate impacts is discussed in more detail in Chapter 7.

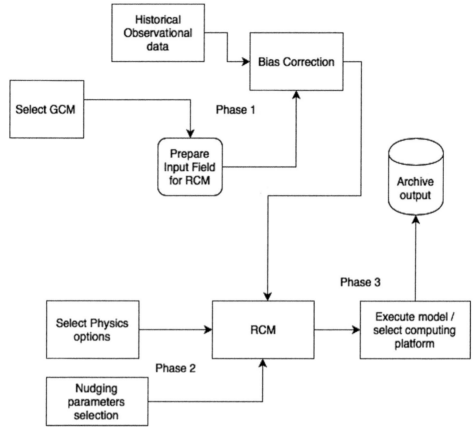

Figure 4.4 Workflow for performing dynamical downscaling calculations. Phase 1 represents the preparation of input data for the model, Phase 2 represents selecting the model physics packages and nudging parameters (optional) and Phase 3 represents running the model to produce the desired output.

4.5 Evaluation of Dynamical Downscaled Model Simulations

Assessing the performance of an RCM is difficult and time-consuming, but usually necessary to determine the value of its results. For example, it is valuable to understand which RCM processes might be driving projected changes, to determine whether or not model outputs make sense. This is especially important when RCMs differ significantly from global model simulations and even disagree with each other.

A wide variety of metrics are used for model evaluations. For example, the root mean square error in model simulations of a specific variable, such as daily temperature, can be calculated for the historical period over a given grid cell or region compared to historical observations (Bruyère et al. 2014). Comparing the

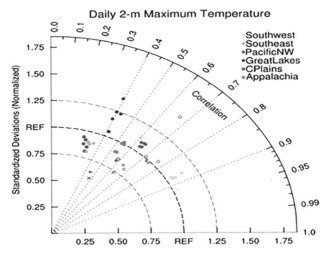

Figure 4.5 Taylor diagram representing the performance of models compared to a reference data set. The X axis represents correlations between observations and model, the Y axis is the standard deviation of the model calculated values from observations, and the lines radiating from the origin represent the root mean square error (RMSE). The point named REF on the X axis and the curve that connects the X axis REF to the Y axis REF is the ideal solution in that it represents reference measurement/observation for model evaluation, has zero standard deviation, RMSE of 1, and correlation of 1. Model simulation results that are closest to this point are better performing than those away from it. For example, in this figure the model output representing the Southwest United States has the best agreement with the reference data, having the highest correlation (~0.9) with reference data and the lowest standard deviation and RMSE compared to other regions.

probability density functions or distributions of selected variables computed from models to those obtained from measurements is frequently used to evaluate the ability of an RCM to reproduce the variability in data. Another commonly used evaluation metric is the Taylor diagram (Taylor 2001), an example of which is shown in Figure 4.5. This is an efficient method for evaluating model performance when compared to observations using three different metrics, namely, the Pearson correlation coefficient, the root-mean-square error (RMSE) error, and the standard deviation, at the same time. Spatiotemporal evaluation methods are useful for comparing the model performance in simulating events that span multiple days and larger regional areas (Wang et al. 2015a).

As discussed in Chapter 3, there are two key components to generating high-resolution climate projections: bias-correction and downscaling. By definition, all RCMs that operate at a finer spatial scale than GCMs are able to downscale. However, RCMs are not equal in their ability to correct for biases, and this ability can also vary within the same RCM by season, by region, and by variable. For that reason, the

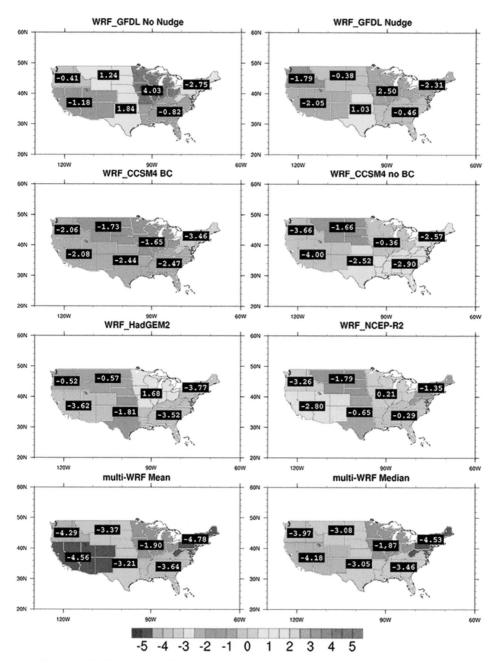

Figure 4.6 Bias in the 95 percent threshold of maximum temperature events between modeled and observed using various model simulations performed with WRF model with various ways of running the model. "WRF_GFDL_nudg" and "WRF_GFDL_No Nudge" are the WRF model with boundary conditions, updated every six hours, that are obtained from a simulation that was archived as a part of the CMIP5 repository by the Geophysical Fluid Dynamics Laboratory (GFDL)

usefulness of the output from a particular RCM for assessing climate impacts and adaptation planning for a given region and application depends on whether the model has a tendency to under- or over-predict variables such as temperature and precipitation when compared to observations for historical simulations. Characterizing RCM by region and by variable is thus a critical step in developing confidence in model outputs and their subsequent use for climate studies (Bukovsky et al. 2013).

RCM bias can be estimated over climatologically or geographically coherent regions within the model domain in most instances (Bruyère et al. 2014; Wang and Kotamarthi 2014). For example, Figure 4.6 shows the precipitation bias, or the difference between observations and model simulations, for the United States from a series of model simulations that were performed using the WRF model. The model domain for these simulations extends over all of North America and simulation is for the ten-year period 1995–2004. Each of the simulations refers to a different GCM used for providing the boundary conditions for the model and different ways of running the model. Some of the simulations used nudged fields, while others used bias-corrected-model initial and boundary conditions. In some of the simulations shown in this figure, the bias between the GCM projections providing the boundary conditions and the observations are first calculated and a correction is applied to selected large-scale atmospheric features to correct for this bias before using these data for running the RCM. It is obvious that some models generally have a larger bias than others, with bias varying from season to season, while other models consistently produce a lower bias across all seasons.

4.6 Availability and Use of Climate Projections from RCMs

A growing body of literature has demonstrated that RCMs add value to the projections generated by the original GCM (Castro et al. 2012; Mearns et al. 2012;

Caption for Figure 4.6 (*cont.*) climate model. "WRF_CCSM4_BC" and "WRF_CCSM4 no BC" refers to simulations that are similar to those described, except that the RCM boundary conditions were from the CESM 1.0 model and they were bias-corrected before use in the WRF model for downscaling for the WRF_CCSM4_BC and no bias correction for the no WRF_CCSM4_no BC case. The WRF_HadGEM2 is the WRF model with boundary conditions from the HadGEM2 GCM with no nudging or bias correction. The simulation labeled WRF_NCEP-R2 is a baseline model simulation that uses the NCEP-R2 reanalysis data to provide boundary conditions to the model. The "multi-WRF mean" and the "multi-WRF Median" are the biases from the ensemble mean and median of the simulations that were performed for this analysis. Each simulation had a spatial grid resolution of 12 km, output was saved every 3 hours, and each simulation was of ten year duration, from 1995 to 2004.
Source: Zobel et al. (2018)

Wang et al. 2015a;), and that these models can reduce model-observational differences or biases in the host climate models, often significantly (Bukovsky et al. 2013; Leung et al. 2013; Torma et al. 2015; Wang et al. 2015a). The question of added value for different downscaling methods is discussed in Chapter 6. Higher-resolution models can also improve the spatial-temporal patterns of change (Di Luca et al. 2012; Wang et al. 2015a).

A challenge from a user's perspective is that, just as in the case of global models, there is no single "best" RCM that can meet all needs. As with global models, regional model intercomparisons illustrate the importance of using multiple regional models to capture a good range of model uncertainty. Recommendations on selecting suitable models for a particular application are provided in Chapter 8.

The most widely available dynamically downscaled output was generated under the North American Regional Scale Climate Change Assessment Project (NARCCAP) using boundary conditions from archived model output generated for an IPCC assessment. The NARCCAP dataset includes one forcing scenario (the mid-high SRES A2 scenario), and multiple regional-scale models at a spatial resolution of 50 km for the periods 1971–2000 and 2041–70 (Mearns et al. 2013). NARCCAP output has already been extensively used in impacts research (see Mearns et al. 2015 for a review).

A global version of the CORDEX is also underway and the studies designed to develop a robust downscaled dataset for many poorly studied regions in Africa and Asia (Giorgi and Gutowski 2015). As a part of this effort, the ongoing North American Coordinated Regional Downscaling Experiment (NA-CORDEX) is performing similar simulations using multiple CMIP5 GCM outputs and multiple RCMs at a spatial resolution of 25 km based on the higher RCP8.5 and lower RCP4.5 scenarios for the time period 1950–2100. Other higher-resolution dynamical-downscaling studies have been performed (e.g., Wang and Kotamarthi 2015) and applied to evaluating climate impacts over North America (Zobel et al. 2017; Zobel et al. 2018), including crop yields (Jin et al. 2017).

A relatively new approach to high-resolution regional climate scenario formation is a hybrid between dynamical downscaling and statistical downscaling (Walton et al. 2015; Sun et al. 2015a). Here, a regional model is run at a very high resolution (e.g., 2 km) with a representative selection of GCMs. Then, statistical relationships are established based on these dynamical results to formulate statistical models for the other GCMs in the set (e.g., all CMIP5 models), thus producing high-resolution results at significantly lower computational cost than dynamically downscaling a large set of GCMs. Walton et al. (2015); Sun et al. (2015a) used five different GCMs in this manner to produce very high-resolution simulations over the Los Angeles, California area. They then constructed a statistical model to approximate the warming patterns for the region using principle component analysis on the five model results on a monthly timescale. The statistical model is a linear combination of the regional mean warming in the GCM and the product of the land-sea contrast with the coastal inland pattern (Walton et al. 2015). The statistical model was validated and then applied to the other CMIP5 GCMs. Thus, this method could provide highly detailed climate scenarios that are deterministically based and represent the full uncertainty of the full suite of CMIP5 GCMs, which is a challenge with computationally expensive dynamic downscaling methods. While this approach has shown promise, there may be some limitations, based, for example, on the spatial extent of the area modeled. It is not clear if the method could function well on a domain covering all of western North America.

As GCMs with variable-resolution capability become more widely available, some of this capability will also become available in GCMs. However, it is likely that RCMs will continue to be a more computationally efficient pathway for producing ensemble runs of downscaled projections and useful for the different applications discussed in Chapter 1.

5

Empirical-Statistical Downscaling

Empirical-statistical downscaling combines observations with global climate model outputs to generate high-resolution spatial and temporal projections. This chapter describes some of the common methods and models used in spatial and temporal disaggregation and bias correction, focusing on aspects of their design and performance that are relevant to their application for quantifying local to regional-scale impacts.

5.1 The Origin of Empirical-Statistical Bias Correction and Downscaling

Empirical-statistical downscaling models (ESDMs) establish a statistical relationship between coarser-resolution model output for a historical period and finer-resolution observed climate variables of interest over the same period. The basis of any statistical downscaling method is a robust record of historical instrumental data that permits calibration at the local scale. The observations used in an ESDM can range from individual weather stations to gridded datasets of surface observations, satellite observations, or reanalysis (meteorological data that have been assimilated into a regular grid using a weather model). The historical period used should be representative of the climatology at that location to ensure the model is able to reproduce observed climate variability. It must be long enough to capture both mean and extreme conditions over many years; thus, at least twenty to thirty years' worth of data are minimally required to build a reliable ESDM for spatial downscaling.

ESDMs are used to bias-correct and/or downscale or disaggregate output from nearly any type of dynamical weather or climate model, including weather forecast models, reanalysis, GCMs, and regional climate models. They do so by combining dynamical model output with historical observations to translate large-scale predictors or patterns into high-resolution projections at the scale of original observations and remove the bias in the model output compared to observations.

ESDMs grew out of the statistical corrections that were first applied to weather prediction models nearly fifty years ago. While the weather models could forecast the evolution of upper-air patterns, they had difficulty resolving small-scale variations in surface weather conditions. Model Output Statistics (MOS) is a statistical technique that uses regression equations to develop relationships between the weather-model output and historical observations that bias-corrected and downscaled the temperature, precipitation, and other local climate indicators simulated by the weather models (Glahn and Lowry, 1972). The need for climate model output at the scale of the proposed impact analysis was first recognized by Gates (1985), in the context of assessing the impacts on ecosystems from climate change, who noted that the statistical relationship between the variations of, say, monthly mean local station data and the corresponding variations of monthly means averaged over an area comparable to that of a GCM grid element could be used to bias-correct the GCM output. Later, a MOS approach was applied to GCM output by Karl et al (1990), who described their approach as: "a method, called climatological projection by *model statistics*, to relate GCM grid-point free-atmosphere statistics, the predictors, to these important local surface observations," and Wigley et al. (1990) who downscaled GCM output to monthly station observations in Oregon. By the late 1990s, the development and application of ESDMs had evolved to the extent that several review papers appeared, summarizing existing methods and their application to impact analyses (Hewitson and Crane 1996; Wilby and Wigley 1997; Xu 1999). As summarized by Wilby and Wigley (1997), by the late 1990s, empirical-statistical downscaling techniques already served "as a means of bridging the gap between what climate modelers are currently able to provide and what impact assessors require."

Over the last two decades, hundreds of peer-reviewed studies have been published describing different methods or additional development of statistical downscaling methods. Mauran and Widmann (2018) provide a comprehensive and up-to-date review. This chapter describes some of the most used statistical techniques and models that have produced high-resolution climate projections available for assessing climate impacts at the regional to local scale.

5.2 Statistical Methods and Models for Bias Correction and Spatial Disaggregation in ESDMs

The statistical techniques used to bias-correct and spatially disaggregate GCM output in ESDMs vary from extremely simple mathematical approaches to sophisticated artificial intelligence and stochastic probability models whose computational costs rival those of RCMs. As discussed, most datasets of high-resolution climate projections that have been generated by ESDMs using methods and models that are relatively simple, with correspondingly low computational costs, in order to generate continuous daily projections through to 2100 based on a wide range of GCMs and scenarios. As describing every advanced method in the scientific literature is beyond the scope of this book, this section focuses primarily on those most commonly used to generate high-resolution projections for impact assessments. Table 5.1 also summarizes various commonly used ESDM models, their downscaling approach and the outputs they produce. For further details on the mathematical origins and development of the ESDM methods and a comprehensive description of their use, the reader is referred to Maraun and Widmann (2018) and Benestad et al. (2008).

The simplest type of ESDM is so simple that people are often surprised to realize it is considered an ESDM: it is the difference or "delta" approach, where a single change or correction factor derived from a GCM is used to shift the entire distribution of a given variable, such as monthly, seasonal or annual temperature or precipitation, up or down over time – e.g., increasing historical observed April temperature by 2.4 °C or decreasing growing-season precipitation by 12 percent (Figure 5.1a). This method can be slightly modified by using a second factor to scale the standard deviation, making it broader or narrower depending on what GCM simulations suggest is likely to happen in the future over that region (Figure 5.1b). Delta methods were used in the original WorldClim database (Hijmans et al. 2005) and the First US National Climate Assessment (USGCRP 2000), and have proven unexpectedly durable and popular, likely because they are so easy to understand and apply. For certain applications, they are still useful: for example, if a given application is primarily interested in shifts in average conditions at the monthly, seasonal, or annual scale, such as changes in mean summer temperature or cumulative annual precipitation, a delta approach is able to produce relatively similar results to a more complex method for significantly less effort (see Section 5.4).

From the beginning, many ESDMs were based on regression techniques where, similar to MOS, multiple large-scale predictor variables are identified that influence the variable of interest. These are then either averaged over an area of interest or reduced to their principle components to create a time series that is then used as a predictor in a canonical correlation analysis (CCA) to relate these to local observations. Regression-based ESDMs have the advantage of employing relatively straightforward and transparent statistics and allowing the user to select multiple predictors that may affect conditions at a given location, including natural modes of variability such as ENSO, upper-air fields, and remote conditions. Predictors must be selected carefully, however; for example, sea level pressure at a given location may be an excellent predictor of day-to-day variability in the historical record; but if sea level pressure does not change significantly as climate changes while another predictor does, the ESDM may not reflect the full

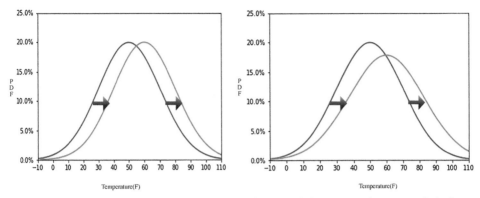

Figure 5.1 A conceptual diagram illustrating the delta approaches to statistical downscaling where (a) a single mean change factor, and (b) a change factor plus a scaling factor derived from a GCM simulation is applied to the observed historical distribution to create future projections at the spatial and temporal scale of the observations.

magnitude of projected changes. The open-source ESDM developed by Benestad et al. (2012) is based on a regression approach, as are many other models in the literature (e.g., Huth 2002; Busuioc et al. 2008; Skourkeas et al. 2013).

If an impact depends on the distribution of daily values, however, a statistical method must be used that is capable of resolving differential changes across the distribution. Due to increasing awareness of the impact of climate change on extremes, and the relevance of information on daily and sub-daily changes to quantifying impacts, many statistical methods used in ESDMs now explicitly resolve, bias-correct, and downscale the quantiles of the distribution. The three main approaches that transform the distributions by quantile are:

- *Empirical quantile mapping*, which calculates the difference between historical GCM simulations and historical observations for each quantile of the distribution and uses that factor to bias-correct each quantile of the GCM distribution from the past through the future (Figure 5.2a). To avoid over-fitting, empirical quantile mapping is generally applied to monthly values, and daily values are then sampled from a historical observed month and adjusted to match the simulated monthly mean (e.g., Maurer et al. 2002; Maurer and Hidalgo 2008). This approach yields daily outputs at the spatial scale of the observational data used, and has been applied to gridded temperature and precipitation over the United States in the BCSD and BCCA datasets originally developed for adjusting GCM output for long-range streamflow forecasting (Wood et al. 2002; VanRheenen et al. 2004), and globally by Sheffield et al. (2006).

- *Parametric quantile mapping*, which can be applied directly to daily values as it first ranks the quantiles from highest to lowest and then fits a statistical function

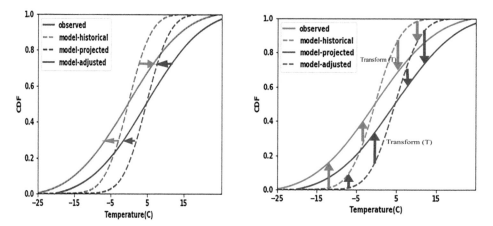

Figure 5.2 These conceptual diagrams illustrate how (a) empirical quantile mapping, and (b) an approach that transforms the CDF bias-correct GCM output using historical data.

to quantile–quantile plot of the observed versus the modeled variable, which avoids the danger of over-fitting (Dettinger et al. 2004; Stoner et al. 2013). This approach generates daily outputs at the spatial scale of the observations and has been applied to both gridded and station-based temperature and precipitation across North America in the asynchronous regional regression model (ARRM) dataset (Hayhoe et al. 2020).

- *Cumulative density functions* (CDF), which, rather than reproducing the relationship between large-scale predictors and local climate, take a probabilistic approach that instead models relationships between their statistical properties (Figure 5.2b). Rather than producing local-scale values, a CDF approach instead produces future distributions. This method is the basis of the CDF-transform or CDF-t method (Michelangeli et al. 2009) that has been used to develop high-resolution projections for a number of locations including Africa and India (Vigaud et al. 2013; Famien et al. 2018).

The idea of bias correcting and downscaling the statistical properties of the distribution rather than individual monthly or daily values lies at the heart of many stochastic methods that are integrated into ESDMs. They can be stand-alone, as the CDF-t method, or be part of a hybrid approach. For example, Stochastic Weather Generators (SWGs) were initially developed to fill in missing data in observations or generating a synthetic long weather record for a given location (Wilks and Wilby 1999). In downscaling, SWGs are often combined with the regression approaches described previously, as regression tends to result in a time series that is less variable than observed, to generate the statistics of daily weather. The most widely used example of a regression + stochastic method is the Statistical DownScaling Model (SDSM, Wilby et al., 2002). SDSM calculates statistical relationships based on multiple linear regression techniques between large-scale (the predictors) and local (the predictand) climate at the monthly scale, then uses stochastic weather generation, with parameters trained on observations, to simulate daily variability at the local scale (Wilby et al. 1998). The SDSM is described as a decision-support tool for assessing local climate change impacts using a robust statistical downscaling technique and was in fact the first downscaling tool freely available to, and easily accessible by, the broader climate change impacts community. SWGs are discussed in more detail in Section 5.4, as they are also one of the primary methods used in temporal disaggregation of rainfall at the sub-daily scale.

It is important to note that all of the methods described so far build a statistical model separately for every location, whether a grid cell or a weather station, and most downscale variables separately, such as temperature and precipitation. The predictor fields used as input to the bias correction and spatial downscaling are physically consistent with each other, being produced by a GCM or RCM that obey physical constraints of conservation of mass and energy for the system being modeled. However, separate treatment of locations and variables in ESDMs has the potential to introduce physical inconsistency to the output, particularly for data points at the tails of the distribution, where the statistical model may differ significantly from one grid point to the next.

While this concern can be addressed by downscaling multiple points simultaneously, or by building multi-variate ESDMs, such methods remain experimental. In terms of available climate projections, concern regarding spatial consistency is addressed by a very different approach to downscaling known as a constructed analogue approach. As described in Hidalgo et al. (2008), this method is "based on the premise that an analogue for a given daily weather pattern from a GCM simulation can be constructed by combining the weather patterns for several days from a library of previously observed patterns" (Figure 5.3). Analog methods are typically applied over large geographic areas and require detailed weather records capable of resolving observed daily weather at relatively fine spatial scale. They have the advantage of ensuring spatial consistency across the domain and can also combine multiple variables such that the selected map, or maps, that form the analogue for a given future day represents a consistent pattern of temperature, humidity, precipitation, and more across that greater region. Analogue approaches form the basis of the MACA (Abatzoglou et al. 2012) and LOCA models, the latter of which was used in the Fourth US National Climate Assessment (Pierce et al. 2014).

One relatively recent development in the efficient, flexible statistical techniques used in ESDMs focuses on understanding and quantifying projected changes to the

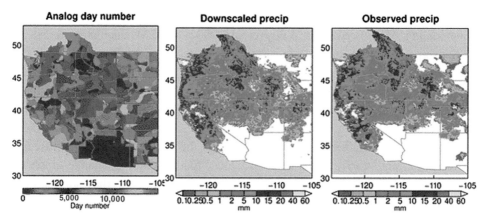

Figure 5.3 A diagram showing the different patterns used to make up an analog in an ESDM.
Source: Pierce et al. (2014)

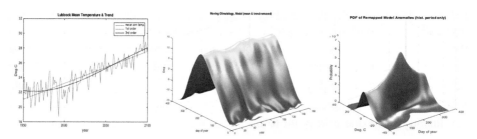

Figure 5.4 New signal decomposition approaches to ESDM reduce the signal to the long-term trend (left), the annual climatology (center) and the daily variability (right).
Source: Hayhoe et al. (2020)

shape of the distribution of one or more variables without having to specify the form of the distribution. This approach is known as "non-parametric," as opposed to parametric approaches, which fit a function (e.g., linear, gaussian, gamma) to the data. For example, McGinnis et al. (2015) found that kernel density estimation (KDE), a non-parametric approach originating in the fields of signal processing and econometrics that allows the user to estimate the probability density or distribution of a variable, was better at removing the bias from model output than parametric distributions, which assume a specific shape to the distribution. A second new development uses signal processing approaches to disaggregate the signal in climate data into individual components – the long-term trend, the seasonal climatology, and the day-to-day variations – that can be analyzed, bias-corrected, and downscaled separately, before being recombined (Figure 5.4). The ESDM component of the Statistical Analysis of Residual Trends (STAR)

Table 5.1 *Summary of Widely Used Statistical Downscaling Methods and Their Typical Characteristics (Source: Kotamarthi et al. 2016)*

Statistical Method	Approach	Geographic Extent	Spatial Resolution	Temporal Resolution	Variables
Delta[a]	Differences between GCM historical and future projections added to historical observations (seasonal, monthly, daily)	Global	From 1/120 to 1/6 degree	Monthly by decade from 2020s to 2080s	Maximum, minimum temperature; precipitation and BIOCLIM variables
Bias correction–spatial desegregation (BCSD) (also sometimes known as BCQM)[b]	Historical data are used to correct monthly GCM output and observations; model bias is empirically corrected for each quantile	Continental United States	1/8 degree	Daily 1950–2099	Maximum, minimum temperature, precipitation; monthly hydrology and wind
Monthly bias correction (MBC), nested bias correction (NBC)[b]	The mean and the standard distribution of the output is corrected with observed precipitation in MBC; NBC also corrects the lag 1 autocorrelations	N/A	N/A	Daily	Maximum, minimum temperature, precipitation
Asynchronous regression[c]	Historical data are used to correct daily GCM output; model bias is removed by a parametric correction for each quantile	Continental United States	1/8 degree and for individual stations	Daily 1960–2099	Maximum, minimum temperature, precipitation, humidity (stations only)
Cumulative distribution function transform (CDFT)[d]	CDF of a climate variable at large scale is related to CDF of the climate variable at local scale	Global	Point/Local	Daily and user-defined timespans	Maximum, minimum temperature, precipitation

Table 5.1 (*cont.*)

Statistical Method	Approach	Geographic Extent	Spatial Resolution	Temporal Resolution	Variables
Equidistant quantile mapping (EDQM)[d]	Uses CDF for historical, model current, and futures to correct the project differences in the CDF between historical and future projections of the model	Global/ Regional	Same resolution as observational gridded data	Daily, monthly, and user-defined timespans	Maximum, minimum temperature, precipitation
Kernel density distribution mapping (KDDM) (McGinnis et al. 2015), currently applied to NARCCAP[d]	Uses a non-parametric transformation to make the estimated probability density function (PDF) of model output match the PDF of observations	Continental United States	50 km and point	Daily	Maximum, minimum temperature, precipitation
Statistical downscaling model (SDSM)[e]	Linear regression between large-scale climate predictors and surface conditions developed using historical data and applied to GCM output	Global	Point/Local	Daily, monthly, and user-defined time spans	Maximum, minimum temperatures, precipitation

[a] *The delta method is the oldest method used to generate higher-resolution information from GCM output. It has been in use since the mid-1980s, and is still in use today. For temperature (mean, maximum, and minimum), the difference between the relevant future time period and the current period is calculated, and then that change is added to observed temperature data. Most often, monthly mean changes are calculated and then added to observed data at monthly down to daily time scales. Thus, only the mean change in temperature is used and only the mean of the future distribution is changed, not its shape (which may affect the probability of extremes at either tail of the distribution). For precipitation, the change is usually calculated as a percentage change in precipitation. The observed precipitation (monthly or daily) is multiplied by the ratio of the future precipitation to the current precipitation from the model. If this ratio is used to modify daily observed data, there are several limitations. First, the frequency of precipitation is not changed, but the variance is (increased if the ratio is greater than 1 and decreased if it is less than 1). An example of application of this method is the SNAP (Scenarios Networks for Alaska-Arctic and Planning) dataset (https://www.snap.uaf.edu/methods/downscaling). See Mearns et al. (2001) for a more detailed view of the delta approach.*

[b] A number of statistical downscaling models incorporate bias correction into their framework. Such methods first use historical observations to correct the bias in the GCM by empirically mapping the monthly distribution of GCM values onto the observed distribution. Spatial disaggregation is then used to achieve finer spatial scales. Some examples of the models that belong to this class are bias correction–spatial disaggregation (BCSD), MBC, and nested bias correction (NBC). The BCSD model developed by Wood et al. (2002) used monthly averaged GCM calculated temperature and precipitation output. The model output is bias-corrected using gridded observations (e.g., Maurer et al. 2002) that are scaled to the grid size of the model. The bias corrected model output is then scaled to the spatial scale of interest, typically the point of observations that are sub-grid scale to the model. A final step performs time disaggregation using daily patterns from observational datasets to scale the monthly projections. The MBC (Johnson and Sharma 2012) is similar in concept to the BCSD, in that the monthly averaged mean and standard deviation of the precipitation from GCM downscaled to an observational location is corrected with observed precipitation. The bias corrections for the future climate are assumed to be the same as the past climate for this method. In addition, the NBC (Johnson and Sharma 2012) bias corrects the monthly mean and standard deviations of the downscaled precipitation, as well as correcting the autocorrelation lag between the present and the next month and correcting an annual precipitation autolag for the present year compared to the next. These methods are commonly applied in hydrological assessments (Sachindra et al. 2014).

[c] Asynchronous regression is similar to BCSD, except that the bias correction is accomplished by ranking observed daily and historical model-simulated daily values by month, then fitting a parametric equation to their quantile–quantile relationship. This approach generalizes the relationship between observations and model simulations to an extent that permits the use of daily, rather than monthly, inputs. A widely used example of this method is the ARRM v1; Stoner et al. 2013), which uses piecewise linear regressions to build monthly q–q relationships based on daily data and, as such, is expected to better represent the tails of the distribution than approaches based on monthly simulations alone. ARRM can be applied to both gridded and individual weather station observations, which permit the downscaling of additional variables including maximum, minimum, and daily average humidity and solar radiation.

[d] CDF-based approaches attempt to downscale the entire statistical distribution of the variable to a local distribution rather than a mean value of a climate variable or its standard deviation. The EDQM method uses the difference between the observed CDF and modeled CDF for the present to calculate a correction that is applied to modify the projected CDF from a GCM (Li Liu and Zuo 2012). The CDFt method is different in that a weather typing scheme is used to correlate a synoptic scale CDF with a locally observed CDF to obtain a downscaled or transformed CDF. As explained by Michelangeli et al. (2009), "the CDFt method assumes that there is a translation available for translating a CDF of a GCM variable (e.g., temperature) to a local scale CDF." The kernel density distribution mapping (KDDM) first develops a PDF of a distribution (temperature from observations and model) and integrates the PDF into the CDF. The transfer function is then built using these modified CDFs (McGinnis et al. 2015).

[e] The Statistical DownScaling Model (Wilby et al. 2002) differs from all those described in that the predictands and predictors are not the same variables. The SDSM uses linear multiple regression to relate large-scale upper-air variables (e.g., 500 mb heights, humidity, vorticity) to the local impact variable of interest (e.g., daily temperature or precipitation, or both). Regression coefficients are determined using reanalysis data for the large-scale variables and point observations for the predictands. These relationships are then applied to future GCM outputs. These techniques generally reproduce well the current point observations, but as with all of the empirical techniques, it is assumed that the relationships between the predictors and predictands do not change with climate change. In addition, the explanation of variance (for the predictands) is not perfect. Typical R^2's for temperature are about 0.8, and are even less for precipitation.

framework (Hayhoe et al. 2020) takes this approach and incorporates KDE to map between the distributions of daily variability generated from observations and GCM output in the STAR-ESDM.

In addition to the methods described, many far more complex and sophisticated methods have been used in ESDMs. Previously independent methods have been combined and additional advanced statistical and computational techniques applied, including artificial neural networks, Bayesian frameworks, support vector machines, clustering methods such as expectation-maximization algorithms, and combined statistical-dynamical approaches as discussed previously in Chapter 4 (e.g., Murphy et al. 2007; Vrac et al. 2007; Tomassetti et al. 2009; Pinto et al. 2010; Walton et al. 2015). Due to their complexity and, in some cases, their computational demand, most sophisticated methods languish unused following their initial publication. However, as Swaminathan et al. (2018) discuss, "many existing machine language algorithms are limited in their ability to model complex functions with many variations that represent deep relationships between input and output variables." Specifically, many off-the-shelf neural networking algorithms only have a few layers; even in deep neural networks, learning algorithms perform poorly when complex functions require errors and uncertainties to be propagated across many levels. Advanced approaches such as convolutional neural networks (CNNs), developed for visual recognition tasks, may offer new avenues for learning based deep multi-layered complex relationships that describe climate processes; these and other approaches remain to be fully explored and implemented in the future.

5.3 Statistical Methods and Models for Temporal Disaggregation in ESDMs

For some applications, temporal disaggregation is required to quantify impacts that depend on higher-resolution information than is available from a model or a dataset. This includes daily, when only monthly is available, or sub-daily, when only daily is available. Applications that may require sub-daily temporal information for temperature include urban electricity demand in large global cities and urban centers across North America, and growing degree-days used to estimate the timing of blooming and fruiting. For precipitation, sub-daily information is essential to informing flood risk. Hence, much of the research that has been done on generating hourly precipitation estimates is in hydraulic engineering and flood risk management related to infrastructure design.

Relatively simple statistical methods can be used to interpolate daily to sub-daily values when these values are known to vary according to a definable statistical distribution function (e.g., a Gaussian or normal distribution) throughout the day, and when the maximum and minimum values are known or observed.

These models have a long history; as Felber et al. (2018) describes, "development of conceptual models relating plant and insect phenology to temperature (as a measure of heat availability) were already being in the middle of the eighteenth century (Allen and Forman1976; Wilson and Barnett 1983)." Variables that vary in this manner include temperature and, to a certain extent, relative humidity. A simple sine or cosine curve or other sinusoidal shape can be fitted to the occurrence of high/low values and the resulting hourly or even sub-hourly values interpolated. While more complex formulae are available, two studies that examined the reliability of such generic inputs found that, even for regions with high topographical information, incorporating site-specific information into the calibration process does not significantly improve performance except at very high elevation sites or if minimum temperature values are not available (Shtiliyanova et al. 2017; Felber et al. 2018).

Derivation of sub-daily precipitation data is significantly more challenging, as it does not typically follow normal distribution throughout the day (though for some regions, precipitation in certain seasons does tend to occur at certain times of the day – e.g., convective precipitation over coastal areas in the late-afternoon and over the Great Plains in summer in early evening). Existing methods for temporal disaggregation of rainfall come primarily come from the field of hydrological modeling and rely heavily on the availability of historical sub-daily precipitation from which to derive representative statistics or samples of the duration and intensity of past precipitation events.

If historical sub-daily precipitation data are available, they can be sampled to create a climatology of simulated sub-daily precipitation events or used to train a stochastic model that can randomly generate events with a climatology consistent with observed. The simplest method is random or stochastic disaggregation, such as used by Jarosińska and Pierzga (2015), whereby 24-hour precipitation is randomly assigned to a given hour. Typically, the user can set a range of "wet hours" from which the random number generator can choose a number and then for each hour select a random amount of precipitation to assign to that hour. Though this method is simple and straightforward, analyses such as that conducted by Sikorska and Seibert (2018) show that, for geographic locations with relatively long effective daily precipitation duration (in their case, about half a day) simple disaggregation methods are more than adequate.

The most common stochastic disaggregation method is known as the Bartlett-Lewis (BL) rectangular pulse process. It forms the basis of the web-based application "Let-It-Rain" described in Kim et al. (2017a) and is implemented in an R package called *HyetosMinute* developed by Kossieris et al. (2018). The latter package allows generation of hourly and sub-hourly precipitation, including a relationship between intensity and duration which improves the statistics. As the BL method has come into common use around the world, additional refinements have been introduced. For example, Kim et al. (2017b) account for the fact that BL treats each parameter as independent, leading to artificially high variability, by using a hierarchical Bayesian model to jointly estimate parameters that consider covariance. Wasko et al. (2015) modified the BL method to include a hierarchical component to represent rainfall characteristics associated with various ENSO phases, where parameters vary as a function of climate state.

Probabilistic approaches such as creating and modifying Intensity-Duration-Frequency (IDF) curves and using SWGs are also frequently used. Approaches based on IDFs are relatively popular, as this is a framework already used within the hydrologic and hydraulic modeling community. Conditional probabilities are estimated based on a range of inputs: high-resolution regional model simulations (So et al. 2017); empirical mode decomposition of existing rainfall intensity of different durations into multiple orthogonal values weighted by probabilities (Adarsh and Reddy 2018b); and theoretical extreme value distributions including Gumbel and log-Pearson II (Subyani and Al-Amri 2015) and gamma distributions (Kaptué et al. 2015). Outputs trained on observations are then often compared to theoretical extreme value distributions to determine which distribution best represents the characteristics of rainfall at that region (e.g., Shao et al. 2016; Syafrina et al. 2018). A recent analysis by Vu et al. (2018) compares the performance of five different SWGs (CLIGEN, ClimGen, LARS-WG, RainSim and WeatherMan) over three different climatic regions. The authors assess the ability of these models to simulate precipitation occurrence, intensity, and wet/dry spells, finding that SWGs that use second-order Markov chain approaches (ClimGen and WeatherMan) are generally best over all three regions, but that specific SWG performance varied by climate type. A complete review of stochastic rainfall generation models is provided by Benoit and Mariethoz (2017).

Finally, a number of statistical resampling techniques have proven popular, beginning with the Method of Fragments (MOF) (e.g., Breinl et al. 2015; Thi and Ball 2015), a non-parametric method that resamples the ratio of the daily to the sub-daily rainfall using a modified nearest k-neighbor algorithm. Random cascade models where branching numbers correspond to time intervals have proven successful at reproducing duration and intensity of extreme rainfall (Muller and Haberlandt 2018). Simulated rainfall timeseries can also be combined with binary sequences of wet and dry conditions to obtain intermittent time series that match observed statistics (Lombardo et al. 2017). For example, Anis and Rode (2015) combine the BL method described previously, as implemented in the Hyetos R package, with a random cascade model to demonstrate how dividing rainfall into four categories before applying the Hyetos package leads to improved results.

Most methods focus on temporal disaggregation of a specific variable; however, there are some generic software packages that apply common and relatively simple methods to a range of variables. For example, the MicroMet package (Liston and Elder 2006) used simple formulations to both spatially and temporally interpolate and/or disaggregate temperature, humidity, wind, solar radiation and precipitation needed for snow modeling. The MELODIST model (Förster et al. 2016) expands on and updates this model, disaggregating temperature using a cosine function whose amplitude is constrained by maximum and minimum temperature, and the time between solar noon and maximum temperature. Sub-daily humidity is based on hourly temperature and maximum and minimum relative humidity values. Wind speed uses a cosine function that can be calibrated for a given site or just a random disaggregation, while shortwave radiation uses the same formula as Liston and Elder (2006), based on time of day, slope, and cloudiness. For precipitation, the user can select from three choices: simply dividing daily precipitation totals by twenty-four hours, using a random cascade method, or using the nearest neighbor information.

The approaches to temporal disaggregation discussed can be applied to both GCM and RCM output. While global model output is more generally available and, as a result, more widely used, many applications also use regional climate model output as available. RCM outputs have the advantage of simulating precipitation at higher spatial scales, and often produce output at higher temporal scales as well. Extremely high-resolution RCM modeling efforts are still experimental but are rapidly becoming more common as computational resources increase.

In summary, despite the increase in level of detail and complexity of temporal disaggregation methods, a number of studies suggest that relatively simple and transparent methods can yield adequate results, particularly when applied to GCM and/or RCM output, where uncertainties in the physical parameterizations of the models and the future scenarios tend to outweigh the uncertainty due to the disaggregation method used. In many cases, in other words, uncertainty in daily precipitation is more likely dominated by the originating GCM and the future scenario than the disaggregation approach used (Alam and Elshorbagy 2015).

Several software packages, from websites to R libraries, have been developed that allow the user to disaggregate multiple variables at the same time using well-documented, tested, and relatively simple methods. Only in the case of extreme precipitation for hydrologic and hydraulic analysis, where the events being considered are on the order of one-in-a-hundred to one-in-a-thousand years, are more complex methods justified, due to the need to quantify uncertainties in the far tails of the distribution.

5.4 Evaluation of Output from ESDMs

Assessing the performance of an ESDM can be relatively straightforward, compared to evaluating an RCM (see Section 4.6). Without understanding exactly what is being assessed, however, such evaluations run the risk of yielding incorrect or even misleading information. This section describes different ways to evaluate EDSM output and explains exactly what each evaluation or comparison method says about the reliability and usefulness of the projections.

First, as with GCM and RCM outputs, comparing ESDM outputs with observations is only valid over climatological periods of twenty to thirty years or more. Each GCM simulation establishes its own internally consistent patterns of natural variability; no day-to-day or even year-to-year correspondence between observations and GCM simulations or their resulting downscaled output should be expected.

In contrast to GCM or RCM output, ESDMs include bias correction so it *is* appropriate to compare observations with the climatological statistics generated by an ESDM for the historical period, such as mean temperature or precipitation variability. When ESDM simulations or output for the training period are compared to observations used to train the ESDM *for that same time period*, the differences that result are a measure of the goodness-of-fit of the ESDM. In other words, it answers the question: to what extent does/do the statistical method(s) used in the ESDM fit the observed data? It should, to a large extent, or it is not accomplishing its purpose. However, a perfect fit is not desired either, as that is often an indication of over-fitting. For example, if a regression model uses too many parameters in creating a proper fit during the training period, it could be fitting the model to the noise or random variability rather than finding the underlying relationship between the observations and GCM output.

Evaluating the difference between ESDM output and observations for an independent evaluation period, a period of time that was not included to train the model, yields more important insights into questions of over-fitting and transferability or generalizability than comparing it to the observations used in training. Transferability or generalizability refers to the ability of an ESDM to reproduce values outside the calibration period; if the model either under- or over-fits the observational data, it will perform poorly. For example, say that sixty years of historical observations are available for a given location. These years can be divided into two sections, one used to train the model and the other to evaluate it and see if it is able to reproduce observations for an independent time period. The division can be straightforward (e.g., use the first 30–40 years to train and the last 20–30 years to evaluate), mixed (e.g., using the odd years to train and the even years to evaluate) or rely on more sophisticated methods such as jackknifed cross-validation (successively leaving one year out and training the model on the remaining years; repeating this exercise for as many years as there are observational data).

ESDM outputs should never be compared with observations from a period of time that combines both training and independent observations, as this conflates a measure of goodness-of-fit with a measure of transferability. Similarly, ESDM output should not be compared with observations from a dataset not used to train the ESDM, because any differences will be dominated by differences between the observational datasets themselves, and these differences can be more appropriately resolved by directly comparing the two sets of observations.

Box 5.1

Case Study: Climate Scenarios for India in the Context of Water Resources

The Indian National Committee on Climate Change (INCC) and Ministry of Water Resources of the Government of India is assessing river basin-scale impacts based on CMIP5 simulations downscaled using a kernel regression-based ESDM (Kannan and Ghosh 2013; Salvi et al. 2013). Salvi et al. (2016) have developed certain experimentation techniques to test the stationarity hypothesis, and they are used for statistical downscaling products for the Indian sub-continent. The downscaled model outputs of the climate variables are now being used in hydrologic models developed for all the river basins in India through the initiative by INCCC, Ministry of Water Resources (Figure 5.5). The primary purpose is to understand the impacts of climate change in terms of different scenarios on the water resources at river basin scale. This assessment will further be used in understanding and analyzing the future water demand and availability to compute possible water stress for different basins. The INCCC projects are ongoing and expected to be completed by 2021.

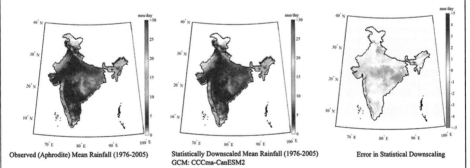

Observed (Aphrodite) Mean Rainfall (1976-2005)   Statistically Downscaled Mean Rainfall (1976-2005) GCM: CCCma-CanESM2   Error in Statistical Downscaling

Figure 5.5 Comparison of observed precipitation, statistically downscaled precipitation, and the difference between the two.
Source: Prof. Sublimal Ghosh, IIT Mumbai, India

Lastly, with some creativity, an ESDM can also be tested for stationarity – the assumption that the relationship between observations and historical simulations is stationary or valid in the future. This type of evaluation is much more challenging, and is typically conducted by researchers rather than practitioners, evaluating the model during development. The first type of evaluation involves using historical data to look for changes in the relationship between large-scale predictors and local climate. Zhang et al. (2011) tested the structural stability of temporally dependent functional observations using long observational records to detect "change points" in the bias of historical GCM temperature simulations relative to observations. Salvi et al. (2016) analyze historical data to carefully select training and evaluation periods with very different conditions, such as cold or warm years, and use these to examine the ability of an ESDM trained on one type of period to reproduce the climate of another.

A second method to evaluate the stationarity of an ESDM is to use output from a higher-resolution GCM or RCM as "truth" and evaluate the extent to which a simpler ESDM is able to reproduce fully dynamic simulations (e.g., Vrac et al. 2007). This method has been codified as a "perfect model" approach by Dixon et al. (2016) who use coarse-resolution GCM simulations to generate future projections, then comparing these with high-resolution GCM simulations for the same future time period. This demonstrates that the assumption of stationarity can vary significantly by ESDM method, by quantile, and by the time scale (daily or monthly) of the GCM input. Even simple ESDMs the stationarity assumption generally is valid in the central part of the distribution for temperature, but more complex methods are needed when downscaling the more extreme tails(Figure 5.6). For all ESDMs tested to date, the stationarity assumption is violated more often in areas of rapidly varying topography (Dixon et al. 2016, Stoner et al., 2017).

5.5 Comparison between ESDMs

As discussed in Section 5.3, there are a wide variety of ESDMs for bias correcting and spatial disaggregation of climate model output. Each of these methods was designed to either serve a specific need or overcome a problem with a given method. Comparing across these various models that use different methodologies is a challenging task. Wilby et al. (1998) performed an evaluation of different methods for downscaling using ESDMs. The models tested included weather generators (WG), perfect prognosis (PP, explained below) and models developed using Artificial Neural networks (ANN). The found the ANN methods performed the worst and there was considerable variation among the methods in predicting precipitation.

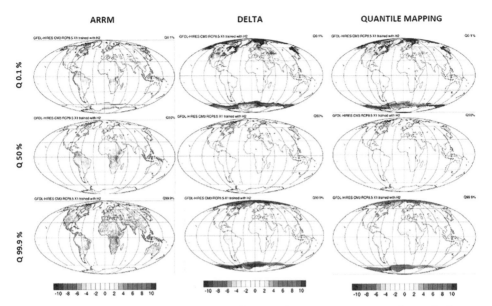

Figure 5.6 These figures show the difference between temperature by end-of-century as simulated by ESDMs (a parametric quantile mapping approach, left; the delta approach, center; an empirical quantile mapping approach, right) compared to a fully dynamic high-resolution global model simulation.

An international initiative, VALUE, was created to compare across a range of ESDMs, create a common community framework and development of tools for evaluating the different aspects of these models (Gutiérrez et al. 2019). This project created a common database of observational dataset from eighty-six weather stations across the countries of European Union. A total of forty-eight ESDMs using a range of techniques and approaches participated in the study.

Figure 5.7 divides the ESDMs that participated in the VALUE study into four categories: RAW, MOS, PP, and WG methods. The RAW method is the output from a reanalysis dataset (ERA-Interim) that was used to create the perfect model baseline data for the project, the MOS method with bias correction, PP models and stochastic WG. The PP methods use the reanalysis data and their correlations with station observations. Examples of these types of methods listed in Table 5.1 are SDSM, CDFt and to some extent KDDM. The MOS methods are the simplest of the entire range of models tested here, essentially a form of the delta correction method with adjustment for bias of the model statistics using observational data (e.g., Gutman et al. 2014). In Table 5.1 the examples of this type of methods are the Delta, BCSD, BCOM, MBC, and NBC. The final technique tested here is the weather generator, discussed in the previous sections in detail (e.g., Wilks and Wilby 1999). Examples of these types of models that

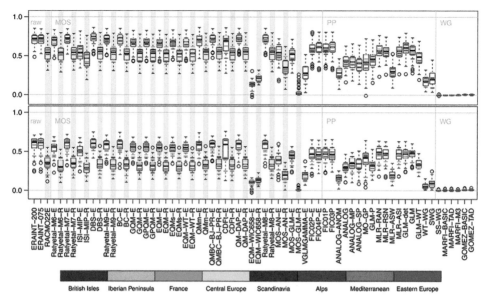

Figure 5.7 Daily correlation of observed precipitation from eighty-six stations with models participating in the VALUE study. The top figure is for winter months (December, January and February) and the bottom figure is for summer months (June, July August). The boxes span the 25–75 percent range and the whiskers extend to the maximum and minimum values, defined as 1.5 times the inter quartile range. Average results for each of the regions shown in the colored label bar is shown in the boxes with the respective color. The x-axis represents the names of different models that participated in the study and y axis is the person correlation coefficient.

Source: Gutiérrez et al. (2019)

are commonly used for downscaling are listed in Table 5.1. Overall, there was no particular approach that is superior and the differences between approaches of a given type of model is larger than the differences between the distinct methodologies uses. One exception is the SWG, as this method uses a stochastic approach for generating weather statistics and is not particularly designed to reproduce observations from a specific station.

5.6 Availability and Use of Climate Projections from ESDMs

ESDMs add value to GCM projections by correcting biases in GCM output such that they match observations (although not always perfectly) over a climatological period and increasing the spatial (and sometimes temporal) scale of the projections. They are best suited for analyses that require a broad range of future projections of standard, near-surface variables such as temperature and precipitation, at the scale of observations that may already be used for planning purposes. However, even within statistical downscaling, selecting an appropriate method for any given study depends on the questions being asked.

As with RCMs and GCMs, there is no single "best" ESDM. Using simpler models can be adequate for certain purposes, particularly those that are primarily interested in changes in seasonal or annual temperature or precipitation. ESDMs that utilize more complex statistical methods are needed to capture nuances in changes in the shape of the distribution of a given variable, particularly at the extremes. A table of available datasets and recommendations on selecting suitable models for a particular application are provided in Chapter 8.

Even though the spatial resolution of GCMs increases with each successive CMIP archive, ESDMs will still be needed for years and even decades to come. New satellite datasets are increasing the resolution of observations available for downscaling and the importance of removing biases from GCM output remains critical for ensuring their use in quantifying impacts at the local to regional scale.

6

Added Value of Downscaling

To what extent does bias correction and downscaling increase the value of GCM outputs for regional-scale applications? This chapter provides an overview of the concept of added value for downscaling studies and discusses the methods and metrics used for evaluating the value that bias correction and downscaling, using an RCM and/or an ESDM, adds to climate projections for impact assessments.

6.1 The Concept of Added Value in Downscaling

GCM projections provide information on broad, regional-scale trends in temperature, precipitation, and other climate variables relevant to assessing the potential impacts of human-induced climate change on human and natural systems. The perception that "higher resolution is better" leads many to assume that the application of an RCM or an ESDM to GCM output increases the value of these projections; however, this is not always the case. The concept of *added value* (AV) helps to clarify this. Specifically, has value been added to the original GCM projections? If so, how? Then, for the user or stakeholder, what is the added value of the high-resolution projections? Do they accomplish something that could not be done without them? Are they reasonable, robust, and trustworthy? To address these questions, this section describes specific situations where downscaling may or may not add value to climate projections, depending on their application.

A key *scientific* question is whether downscaling GCM outputs increases the accuracy as well as the precision of the projections. If the former, then there is added value as the resulting projections are more accurate. If the latter, there is not necessarily added value. For example, higher-resolution spatial information can be obtained by simply re-gridding GCM output through spatial interpolation to any resolution. Similarly, higher-resolution temporal information can be obtained from simply interpolating between individual days or months, or using a weather generator to produce statistics that may match the historical period but have no information on how they are changing in the future. But precision is not the same as accuracy: if GCM grid cells of 10,000 km^2 were re-gridded to finer grid cells of 100 km^2 or if daily values were interpolated to hourly without introducing any new information, while the result would be precise to 100 km^2 or an hour, the information would only be accurate at the same spatial and/or temporal scale as the original GCM.

Does this exercise provide added value? It may, under some specific conditions. First, the impact analysis (a landscape model, a hydrology model, a crop model, or other) must require high-resolution input. However, that input, and how it changes over time, must not vary significantly over spatial scales finer than that of the original GCM, and the analysis must be robust even if the spatial accuracy (rather than the precision) is taken to be that of the GCM. These are rare conditions, and it is difficult to imagine a terrestrial location where this would be the case. However, without understanding how a high-resolution gridded dataset was created, it is difficult to ascertain whether there is added value to using it or not.

For those who design, evaluate, and apply empirical-statistical and dynamical downscaling models, the concept of added value takes on a more specific definition. In broad terms, it addresses the question of whether the downscaling provides more useful, credible, robust, and detailed information than what would be obtained from the GCM alone.

A key *application* question is whether the downscaling adds information that is useful to the user (Rummukainen 2016). For example, a hydraulic engineer may require information on how sub-daily or hourly precipitation is projected to change in order to incorporate climate projections into the design and construction of key infrastructure. Downscaling of daily GCM output could produce daily outputs, and these outputs could be stochastically disaggregated into hourly values; but unless the disaggregation process includes information on how the distribution of sub-daily precipitation is likely to change, the resulting projections can only offer insight into how the daily values will change; they have nothing further to add about how the distribution of sub-daily precipitation is likely to shift in the future. However, if the downscaling process involves the use of a convection-resolving RCM that has been shown to reduce the bias of the precipitation simulated by a dynamical model compared to observations, and can generate sub-daily output, valuable information has been added to the analysis and, for the hydraulic engineer, there is added value in the downscaling process.

From both a scientific and an application perspective, a third key question in determining added value relates to uncertainty. What if uncertainty is unknown, or not computable using the current state-of-the art models or observational datasets? For example, there is some indication that, as climate changes, conditions conducive to large tornado outbreaks are increasing. However, the scientific ability to simulate accurately the response of these weather events to human-induced change over time scales of decades to centuries is still in its infancy. Thus, providing information on future tornado risks may be either impossible or so uncertain that it does not add value to the analysis. Similarly, a given human or natural system may be so complex or so poorly understood that it is impossible to predict how it will respond to future climate changes. Even with well-quantified climate projections for that region, current understanding of the system itself may be insufficient to apply this information; thus, climate projections – whether downscaled or not – are unable to add any value to any decisions currently being made with regards to that system.

Finally, it is not uncommon for the introduction of new information to result in a broader rather than a narrower range of uncertainty. Does this mean there is no added value? From a scientific perspective, we would argue that the opposite is the case. Planning for the future involves an accurate assessment of risk, and if the risk envelope is wider than previously thought, this is important information that should be factored in.

6.2 Added Value from the Perspective of Scientists and Decision Makers

Added value can be determined from a scientific point of view, when downscaling provides information that is judged to be more robust and credible than that derived from a GCM, and also from a user point of view, when that information allows decision makers to quantify impacts and/or inform resilience, response, or adaptation planning in a way that they would not have been able to with GCM output alone. Often, added value will mean the same thing for the scientists and the users, as adding information to climate projections will ultimately result in value for an end user.

Thus, from the perspective of climate scientists, added value is simply the idea that performing downscaling using an RCM or an EDSM: (a) produces output that is more accurate and less biased when compared to observations; and (b) either introduces new information to a downscaled variable that is not resolved at the spatial scale of a GCM or resolves a phenomenon that is not represented in a GCM. The most frequently cited way that downscaling adds value to climate projections is through resolving the effects of topographical features that occur below the grid scale of a typical GCM, such as the influence of large lakes on neighboring regions or of the ocean on coastal climate. Compared to observations, most downscaling methods can introduce value as a result of resolving these features on regional scale even when these results are averaged back up to a grid scale of GCM (Wang et al. 2015b).

From the perspective of a user of this information, added value may lie in the customization, translation, and/or localization of information. For example, global temperature has increased 1°C over the last century; but for the city of Chicago it is more useful to know that, locally, seasonal temperatures have increased by 1–2 °C and the greatest changes have occurred in winter. Even seasonal changes may not be as useful as information on specific types of events, such as the number of days below freezing, or the number of days per year above a commonly-used measure of extreme heat, such as 35 °C or 100 °F. Geo-locating observed and projected future changes to a high-resolution grid or individual weather stations may help a city or a region with high spatial diversity in mapping the projected impacts, especially for events such as heat waves that disproportionately affect areas with a high urban heat-island effect (such as neighborhoods lacking tree cover) and those of lower socioeconomic status who are more vulnerable to these events. Producing this type of information would involve some method of downscaling, either using an RCM or an EDSM, capable of producing output at the spatial and time scales required for the application.

For the user, there may be value in the ability of downscaling to generate projections that can be used as input to modeling and decision-making frameworks already used in, for example, ecosystem or water management or crop-yield calculations. Downscaling and translating GCM output into information that can be read directly into a crop model, a watershed hydrology model, or even an epidemiological model enables experts in those areas to quantify potential impacts of climate change directly on metrics relevant to their sector. An example would be to use downscaling to create and prepare the temperature, precipitation, humidity, and solar radiation inputs required for a water management model that would then be used to inform water allocation strategies for the next few decades. This is not added value in the sense described earlier, but translating climate-model results to formats that are useful to an end user does provide additional value that is not directly available from GCM output.

The concept of added value is an important one, and one that must be considered at the beginning of the application, in order to identify the most appropriate source(s) of information as well as the intended goals of the project.

6.3 Added Value in the Context of Dynamical Downscaling

The regional climate modeling research community has devoted significant effort to determining and quantifying the added value of their efforts. A great deal of research focuses on determining whether a higher-resolution regional model better reproduces the current regional climate than the driving GCM (IPCC 2013) and, on a process level, whether it produces credible changes under future forcing. As discussed, RCMs accomplish this by adding information on topography and physical processes that is not directly resolved in GCM simulations nor producible by simply interpolating GCM outputs to a higher spatial or temporal resolution.

Thus, added value in RCMs is typically associated with finer scale phenomena such as cyclones and storms, and environments where higher-resolution forcing come into play, such as in mountainous areas, complex coastlines, and inland bodies of water. For example, Torma et al. (2015) explored how different spatial resolutions affected RCM ability to reproduce observed precipitation over the Alps and found that higher-resolution RCM simulations were most successful (Figure 6.1). They also demonstrated that RCM outputs had significant added value for precipitation extremes, an important metric for impacts analyses such as flooding. In another example, Wang et al. (2015a) showed that higher-resolution models are better able to reproduce the observed spatiotemporal correlations of precipitation across a region compared to the GCM, as the spatial extent of precipitation is correlated through large scale precipitation events (Figure 6.2).

A relatively recent and important advance in added value is the fact that, as the resolution of typical GCM and RCM simulations has increased, many RCMs are now able to be run at a resolution of 4 km or higher. This enables an RCM to resolve directly, rather than parameterize, convection. The added value of convection-permitting RCM simulations relative to simulations from coarser-resolution RCMs primarily occurs in areas where deep convection is an important process, such as the tropics, as well as in mountainous regions and for extreme events including storms and hurricanes (Prein et al. 2015). However, simulations of such high resolution require considerable computer resources, so the multi-decadal simulations from convection-permitting RCMs that are needed for impact assessments are still relatively rare at this point in time.

One of the key issues regarding added value for regional climate modeling is how to measure it. For comparing observations with historical RCM simulations, several useful metrics have been developed. For example, an added value index (AVI) has been developed by Kanamitsu and de Haan (2011), based on a characteristic spatial distribution of skill rather than average values for regional models; a second potential added value index (POV) was developed by Di Luca et al. (2012), using variance decomposition techniques.

Determining added value is more difficult in the context of projections of climate change. Since future observations are not available to be compared with future simulations, there are no "metrics" for the future. Instead, the added value must be demonstrated in other more process-oriented ways, such as examining what happens to the relevant physical mechanisms governing the spatial details of climate and comparing how these processes change in the GCM vs. the RCM simulations (Paeth and Manning 2013). This type of process-level analysis is often referred to as "credibility analysis" – i.e. how credible are the projections? (e.g., Barsugli et al. 2013; Bukovsky et al. 2015; Mearns et al. 2017).

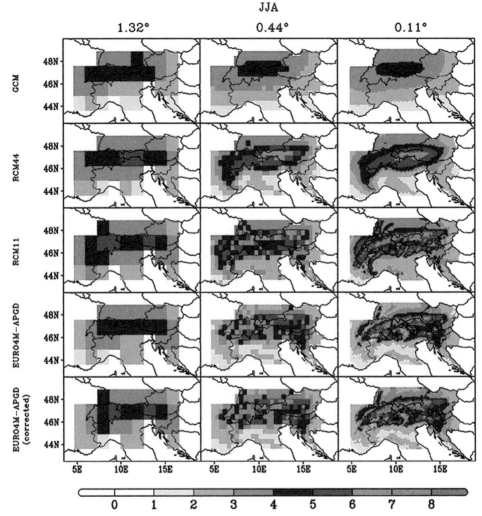

Figure 6.1 Comparing summer precipitation over the Alps for the historical period 1976–2005 as simulated by global climate models (first column), downscaled by a coarser-resolution RCM (second column), and downscaled by a finer-resolution RCM (third row) with Euro4m-APGD observations (fourth row) shows how higher-resolution RCMs are better able to simulate climate in regions with fine-scale topographical variations, even when the fine-scale results (right column) are upscaled back to the resolution of the coarser RCM (center column) or the driving GCM (left column).
Source: Torma et al. (2015)

Over Europe, for example, Giorgi et al. (2016) demonstrated that climate projections downscaled by RCMs with a spatial resolution of 12 km can even alter the *sign* of projected changes in precipitation relative to the driving GCMs. Specifically, analyzing projected changes in future summer precipitation, they

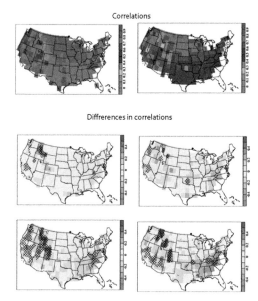

Figure 6.2 Comparing the observed correlation in observed monthly precipitation (top panel) for locations that are 2 degrees apart in latitude (left) and 4 degrees apart in latitude (right) with high-resolution simulations (middle row) and coarser-resolution simulations (last row) shows that the fine resolution model is closer to observations, adding value by improving the correlations in the spatiotemporal fields of precipitation over most of the CONUS and over the Western United States in particular.
Source: Wang et al. (2015a)

found that the RCMs produced significant precipitation increases at higher elevations in the Alps. They then demonstrated that this increase, which is not simulated by the GCMs, is the result of a credible physical mechanism: in this case, increased convective rainfall, which results from an increase in potential instability that in turn is caused by high-elevation surface heating and moistening in a warmer climate. Another analysis by Evans and McCabe (2013) for the Murray–Darling basin in Australia showed that, as the resolution of RCM simulations increased, simulated patterns of both temperature and precipitation change gained both detail (particularly for coastal and mountainous regions) and credibility compared to the driving GCMs. These are only two of many examples where higher-resolution RCM simulations of future climate have been shown to differ in important and often more credible ways from the projections generated by coarser-scale GCMs (e.g., Bukovsky et al. 2015; 2017; Lee et al. 2019).

RCM simulations can also reveal uncertainties in future projections that may not be obvious from GCM simulations alone. In some cases, downscaling narrows the range of uncertainty in future projections compared to a GCM, but in other cases it can actually broaden it in a way that is meaningful and better represents the physical processes involved (e.g., Mearns et al. 2013). For example, even for a variable as basic as mean annual temperature change by the end of century, some regional models will run hotter than the GCM and others cooler, even if all are using the same GCM as input (e.g., Bukovsky et al. 2019). This suggests the presence of higher-resolution physical processes that may amplify or dampen the influence of broad-scale spatial trends on local change.

In summary, a number of studies have reviewed the question of added value in regional model simulations of future climate (Feser et al. 2011; Di Luca and Laprise 2015; Wang et al. 2015b; Rummukainen 2016). While the added value can vary by location, variable, and even the time of year, these reviews conclude that RCMs do generally add value compared to GCMs (Bukovsky et al. 2015). The one general exception is the fact that the added value provided by the RCM is still partially dependent on the boundary conditions from the GCM, so poor-quality boundary conditions from driving GCMs can result in negligible or "useless" added value – i.e., garbage in, garbage out. For example, Mearns et al. (2017) demonstrated for the southern Great Plains of the United States that high-resolution RCM simulations driven by a particular GCM added no value since the boundary conditions from that GCM were fatally flawed. This could instead be termed "false" added value.

6.4 Added Value in the Context of ESDM

Compared to the RCM community, added value is not the subject of much research in the ESDM community. This is due to the fact that ESDMs, being trained on historical observations, by definition reproduce those observations to the extent that the statistical methods used in the ESDMs are able. But, as discussed, added value is not solely a function of accurately reproducing observations. As discussed previously in Chapter 5, comparing ESDM simulations to observations for the training period simply illustrates the goodness-of-fit of the model, while comparing ESDM simulations to observations from the same dataset but a different time period quantifies the transferability of the ESDM, or the extent to which the ESDMs are able to replicate observed climate conditions not used to train the model. This latter comparison also reflects the adequacy of the training data; i.e., did the initial sample adequately capture a sufficient range of climate conditions? In this sense, it is similar but not identical to historical RCM simulations being compared to observations to determine their added value.

Although the term "added value" is not typically used, evaluating ESDM performance using an independent historical validation period is a common approach when developing a new model or publishing a new dataset. In Figure 6.3, ARRM, a parametric quantile mapping approach, is evaluated relative

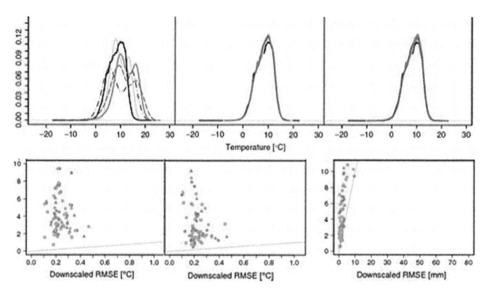

Figure 6.3 (Top) Probability distributions for daily minimum temperature for Half Moon Bay, CA showing 1960–2009 observed daily minimum temperature (black) and GCM-simulated values (left column), downscaled for the training period (middle column), and downscaled for an independent validation period (right column) show that the ARRM model is transferrable – i.e. able to reproduce observations for an independent validation period. (bottom) The change in RMSE compared to observations for GCM simulations (y axis) compared to ARRM-downscaled simulations (x-axis) for maximum temperature (left), minimum temperature (center) and precipitation (right) shows that RMSE is significantly reduced by downscaling.
Source: Stoner et al. (2013)

to observations from station data. The top panel shows that the model's ability to reproduce observed temperature for the training period vs. for an independent validation period is nearly identical, while the bottom panel shows how the root mean square error of GCM-simulated maximum and minimum temperature and precipitation at twenty different weather stations compared to observations is significantly reduced when the GCM outputs are downscaled. As discussed in Chapter 5, the Multivariate Adapted Constructed Analogs (MACA) ESDM uses maps rather than individual time series to downscale GCM projections; comparing the results to observations for the same time period as the maps used during training demonstrated that it is able to improve GCM simulation of temperature, humidity, wind speed, and precipitation over the western US better than simple interpolation would (Abatzoglou and Brown 2012). Finally, regarding the application of this information to quantify impacts, an analysis of streamflow forecasting in Quebec found that downscaling added value, compared to the use of direct numerical weather prediction model output, in both precipitation occurrence and amount that propagated through to improve overall flow forecasts (Muluye 2011).

As with RCM simulations, determining added value is more difficult for future projections of climate change. Again, future observations are not available to be compared with future downscaled simulations. In the case of ESDMs, however, it is not typically possible to conduct the type of process-based analyses discussed in Section 7.3, either. So how can the credibility and – for ESDMs – the stationarity of the models be assessed for the future? Chapter 5 discusses a number of creative approaches to evaluating ESDM stationarity, including the "perfect model" approach which uses high-resolution dynamical simulations from a global model as "future observations" and evaluates the extent to which an ESDM trained on coarse GCM simulations is able to reproduce the regional features of projected climate change at a significantly lower computational cost than running the high-resolution global model.

As implemented by Dixon et al. (2016), the "perfect model" approach uses high-resolution 25 km output from the GFDL-HiRAM-C360 model as "observations" for both past and future periods and coarsened 200 km versions of the same fields as "models." They furthermore define "added value" as occurring if the mean absolute error of the downscaled future output (the downscaled value minus the high-resolution value) is less than the difference between the coarse-resolution value minus the high-resolution value. Applying this approach to the ARRM ESDM in Figure 6.3 demonstrates that there is added value (i.e. the difference between the downscaled versus the high-resolution value is less than 1.5 degrees) across the continental United States except for very high elevation and coastal areas.

Extending this "perfect model" approach to multiple ESDMs yields valuable insights into their relative added value (Figure 6.4). For temperature, for example, nearly all ESDMs produce comparable results for downscaling mean seasonal values. For both hot and cold extremes, however, methods that do not resolve the observed and projected future distribution of daily values produce significant biases, particularly at higher latitudes. For precipitation, a simple delta approach produces large biases across the entire distribution (with the exception of the western United States) that become largely negative by end of century. For projected changes in precipitation values below the ninetieth quantile, both daily and monthly quantile mapping and quantile regression perform adequately. For high precipitation extremes, monthly quantile mapping produces large positive biases relative to the high-resolution dynamical model simulations; only daily quantile regression produces significantly lower biases. These results show clear limitations on simpler methods when simulating projected changes in extreme temperature and precipitation as compared to high-resolution dynamical simulations of climate change at the regional scale.

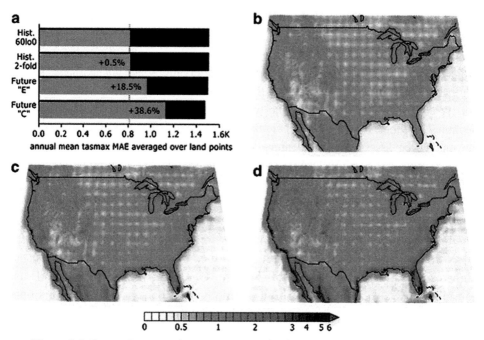

Figure 6.4 Comparing annual-average mean absolute errors (MAE) calculated for the ARRM method's downscaling of maximum temperature for a historical simulation (top right) with two different future simulations (bottom) shows that for most regions, the MAE is less than the MAE obtained from the coarse model relative to the high-resolution model.

Source: Dixon et al. (2016)

Interestingly, the perfect model approach can also be used to inform more detailed, process-level evaluations more similar to what is conducted with RCMs. For example, applying both a parametric quantile mapping (ARRM) and a non-parametric distribution mapping method (STAR-ESDM) to downscaling temperature over high-elevation locations such as the Himalayas showed a large and significant bias in maximum temperature over these locations (Figure 6.5). Further analysis showed that this bias was occurring primarily in one month of the year: July in the Himalayas. Comparing this date with that of maximum streamflow and snowmelt in the Hindu Kush revealed that this bias was the result of the fact that the ESDMs were not accounting for a shift in the timing of snowmelt in the future, which in turn affects local albedo and hence temperature. Understanding the process responsible for this bias constrains the conditions under which ESDMs offer added value, but also provides valuable insights that could be incorporated into future ESDM development to account for this factor.

A process-level analysis of the ESDM suite that participated in the VALUE study (discussed in Chapter 5; Gutiérrez et al. 2019) also provides some insights

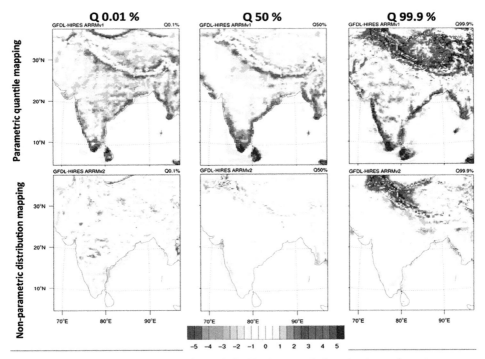

Figure 6.5 Comparing the bias in statistically-downscaled projections of maximum daily temperature on the one-in-a-thousand coldest day (left), average day (center) and one-in-a-thousand hottest day (right) for the 2090s compared to values simulated by a high-resolution global model shows that the bias, and hence added value, depends on both location and month. The top row shows results from a parametric quantile mapping ESDM (ARRM) while the bottom shows results from a non-parametric (i.e. probabilistic) distribution mapping ESDM (STAR-ESDM). Red indicates a positive bias, where downscaled values are too high. The values shown in this figure are annual, but when plotted for July only, the bias over the high-elevation locations in the Himalayas is identical. This suggests it is directly connected to a physical process that is unique to that month, namely peak snowmelt which affects the albedo and the evaporative processes in the region.
Source: Stoner et al. (2013)

into the ESDMs and their ability to represent process scale phenomena. Figure 6.6 shows the number of intense moisture advection events over the UK caused by a low-pressure system that appears over the west of Ireland (Soares et al. 2019). This is an example of a process that causes local or regional phenomena that can be correlated local and regional meteorological features. When these events occur, they bring strong moisture advection over southeastern part of England and the precipitation is typically enhanced by factors of 2–3 on the following days. A positive bias in this metric means that precipitation is overestimated and a negative bias indicates lower precipitation. The models that participated in the

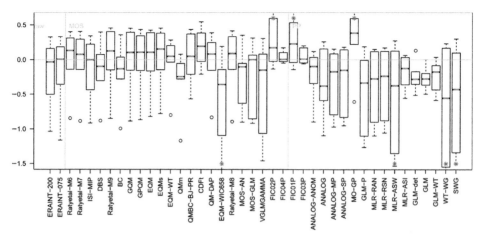

Figure 6.6 Process scale evaluation of ESDMs participating in the VALUE model evaluation projects (Gutiérrez et al. 2019). The process evaluated in this figure is the increase in moisture over England that accompanies a low-pressure system to the west of Ireland. The figure y axis is the anomaly in the number of positive event counts compared to no event counts and the x-axis is the various models that participated in the study.

Source: Soares et al. (2019)

VALUE study as described in Chapter 5 are divided into RAW, MOS and PP. The RAW results indicate that the reanalysis fields used for creating the perfect model for the case study has the tendency to slightly underestimate the event count. The MOS models generally overestimate the phenomena, while the PP models underestimate the response of precipitation to this influx of moisture. There are a few models that perform better than others in both MOS and PP categories, suggesting future research to determine the reason why some models perform better than others while using the same methodology.

6.5 Comparing Statistical and Dynamical Downscaling

In selecting an appropriate downscaling method for an intended application or purpose, it is important to consider the nature of the downscaling method used. In reality, however, it is difficult to do this in a comprehensive and thorough manner. There is no standardized framework, look-up table, or other type of metrics that would make it easy to identify which methods or models were appropriate for which applications, although there are studies that assess the added value for RCMs and ESDMs in general, as well as for specific models. This is of particular importance from a stakeholders' and impacts-analysis point of view, since there are a large number of prospective data sets available providing downscaled data, but few cross data set comparisons or deep analysis, making it difficult for stakeholders to sensibly choose among the available options. The bottom line is that context matters: it affects the credibility and appropriateness and usefulness of methods.

In addition, most studies evaluate only RCMs or only ESDMs; but it is particularly important to evaluate both dynamical and statistical downscaling methods together. A number of basic statistical comparisons have been produced (e.g., Mearns et al. 1999; Wilby and Wigley 2000; Hay and Clark 2003; Wood et al. 2004; Spak et al. 2007; Gutmann et al. 2012; Vavrus and Behnke 2014; Tang et al. 2016; Manzanas et al. 2018), though deeper process-level comparisons are rare. Most of these superficial comparisons show that the results of dynamical versus statistical downscaling differ, particularly for precipitation, but they do not provide deep analysis of why. For example, Tang et al. (2016) applied the SDSM statistical model to 550 weather stations throughout China using input from two GCMs, for a twenty-year historical period and a mid-twenty-first century period. They then used the WRF regional model at 50 km resolution over East Asia using boundary conditions from the same two GCMs to simulate climate conditions over the same two periods of time. Comparing statistical and dynamical model results indicated some differences in projected changes in temperature, but substantial contrasts in changes in precipitation for model results from both GCMs. For example, wintertime precipitation is projected by SDSM-ECHAM5 to increase by more than 50 percent over the Yangtze River valley, while a drying trend is projected by the WRF-ECHAM5 projection (Figure 6.7). The authors suggest that complex factors involved in the precipitation process and the relative few predictors considered in the statistical downscaling account for the differences in the changes in precipitation.

Some more sophisticated analyses have been performed to establish why statistical and dynamical downscaling results often differ (Sun et al. 2019). They compared WRF dynamical simulations with two sets of LOCA statistical projections over California, one of which was trained on typical gridded observations and another trained on historical WRF output. This enabled a separation of projection differences into two categories: differences in training data (i.e. LOCA trained on WRF vs. gridded observations) and differences in how WRF and LOCA handle climate change signals. Focusing on future changes in springtime warming over the Sierra Nevada, results indicate that only WRF was able to capture accurately the snow albedo feedback in its temperature patterns: the same point made by the "perfect model" comparison shown in Figure 6.5. The importance of surface feedbacks was also demonstrated in Mearns et al. (1999).

For both ESDMs and RCMs, it is possible but challenging to conduct a process-level assessment of what the models do and do not include, and thus how

Summer Winter Annual

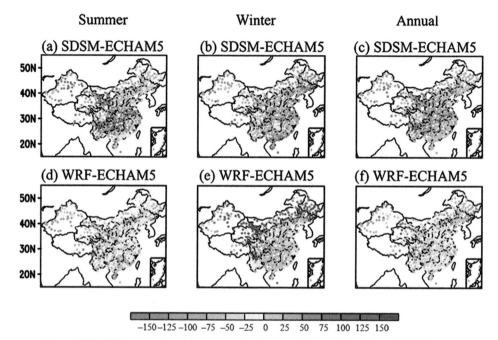

Figure 6.7 Differences in the projected downscaled precipitation fields from an EDSM and an RCM using the same GCM inputs. The EDSM used for this study was the Statistical Downscaling Model (SDSM, Wilby et al. 2012) and the RSM is the WRF model. Downscaled precipitation fields from ECHAM5 (Jungclaus et al. 2006) climate model are compared. The ECHAM5 results are from CMIP3 for the SRES A1B scenario. The figure shows percentage changes in precipitation for 2041–60 compared to 1981–2000. In some locations in Western China the precipitation changes estimated are opposite in sign and the values in general between the two methods are different by 20–30 percent.

appropriate they may be for a given application. For example, as discussed in Chapter 4, in the area of regional climate modeling, models with horizontal spatial resolutions finer than 4 km are referred to as "convection-resolving models" as they are able to explicitly rather than implicitly simulate convection. If the application is one that requires, for example, diurnal precipitation rates from organized mesoscale convective systems for infrastructure applications, then this is the resolution of RCM that should be used (see Section 7.3). Of course, that is easier said than done, because currently RCM simulations at this resolution are typically conducted for scientific research purposes rather than impact studies. However, the field is changing rapidly and as more computational power becomes available, such higher-resolution simulations will become the norm. In empirical-statistical downscaling, many models do not resolve changes in sub-monthly or sub-daily variability. For an application where downbursts or heavy precipitation events are responsible for most of the impacts, it is important to have some sense of how these statistics may be changing. Conversely, if the ESDM does not simulate changes in these metrics, it may not be relevant.

6.6 Research Needs to Further Determine Appropriate Use of Different Methods

Information on future climate and how it will differ from what has been experienced in the past is essential to inform human decisions across a broad range of sectors and regions. To both assess the outcomes and guide in the selection of our choices today and in the near future, we need information about what the result of those choices may be. In some cases, even the best available science can simply say that the risks are real. In other cases, it is possible to state that the risks are changing: getting better or worse, shifting seasonality or intensity or frequency. Here, downscaling will find a use. In others, it is possible to quantify projected change and its uncertainty: in drought, heavy rain events, extreme heat, and more; and downscaling plays a key role in generating this information.

The ideal "added value" of high-resolution climate information is simply to produce credible information that improves the impact assessments and analysis, understanding, and decision making by providing key insights needed to make resilient, climate-ready adaptation and mitigation decisions today in order to minimize present and future risk.

For some of these risks, the science cannot yet provide actionable information. For other risks, the decision-making structure is not well enough understood or defined, or the vulnerability of the system is not clear. But for some, information can be provided that alters the trajectory of decision making and planning compared to if it had been based on historical climate data alone. Generating and applying actionable information is at the innovative edge of combining the physical, social, and natural sciences, combining art with science and learning through the experience of doing so.

Box 6.1
The Infrastructure and Climate Network (ICNet)

There are many sectors and communities who are already using downscaled GCM outputs and understand the added value of downscaling. For the transportation sector, which spends billions and billions of dollars on infrastructure every year, extreme events, as well as more subtle and pervasive changes in long-term temperature and precipitation regimes from climate change, can interrupt the use of transportation

Continued

Box 6.1 (cont.)

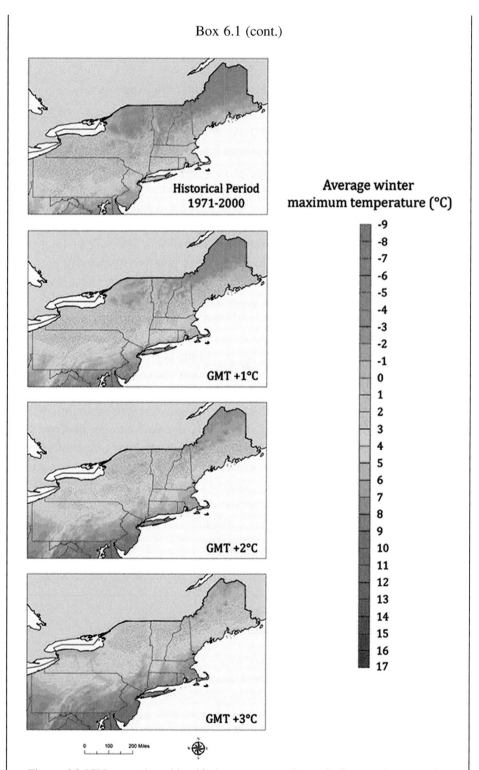

Figure 6.8 ICNet members identified twenty-one climate indicators that are relevant to the transportation infrastructure community. The ICNet used ESDM

Box 6.1 (cont.)

infrastructure, increase maintenance and repair costs, and accelerate infrastructure deterioration process.

The Infrastructure & Climate Network (ICNet) members, embodying over a hundred climate scientists, civil engineering researchers, engineering practitioners, social scientists and policy makers, are collaborating to accelerate research on climate change impacts on road and bridge design in the United States. In the Northeastern United States, like many regions, the changes in climate that occur over relatively small distances dramatically change how the transportation network is designed and maintained. The added value of downscaled GCM outputs is helping the ICNet to address a range of questions: What is the role of climate in infrastructure design? What hinders the inclusion of emerging climate and sea level rise information in engineering design? What is the most critical information required for the engineers to design for a changing climate?

For example, the region's "low-volume" roads, roads with traffic volumes generally less than 400 vehicles per day, are highly sensitive to winter temperatures. They often provide access to forests for logging and other forest-management activities. Seasonal freeze–thaw conditions control the ability of vehicles to use these roads safely. In the coldest winter periods, heavier vehicles are sometimes permitted because the road has increased strength and stiffness. However, during spring-thaw periods, frozen roads begin to weaken as they thaw, becoming subject to damage from heavy trucks, or even entirely unusable. Allowing trucks to haul heavier loads during the winter benefits the forestry industry. Lower weight limits during the spring thaw protect roads and reduce repair costs. GCM outputs indicate that these recent warmer and shorter winters will become increasingly warm and short in the future. Downscaled GCM output showed that northern New Hampshire's White Mountains region, whose frozen road season currently averages about twelve weeks per year, will experience a decrease of nearly three weeks by mid-century, but that coastal roads' nine-week frozen road season could routinely disappear by mid-century (Figure 6.8). This difference in information allows state transportation agencies and foresters to make policy, practice, and major capital purchase decisions with an eye toward the future.

Caption for Figure 6.8 (*cont.*) methods to create a series of maps to illustrate the projected changes in these precipitation and temperature conditions in the northeast United States as global mean temperature (GMT) rises. The example shown provides the average winter maximum temperature for the historical conditions (1971–2000) alongside a map of projected future conditions that will result from successive increases in GMT (+1 °C, +2 °C, and +3 °C, or 1.8, 3.6, and 5.4 °F, respectively).

Continued

Box 6.1 (cont.)

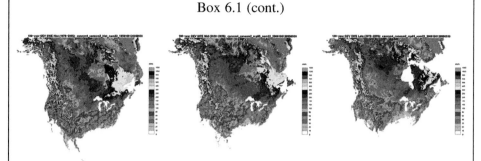

Figure 6.9 Engineering design maps created for 100-year annual maximum snow water equivalent across North America for historical (1976–2005), mid-century (2030–59), and late-century (2070–99). The maps were created using dynamical downscaling methods to illustrate the projected changes in snowpack as GMT rises.

Winter snowpack and associated snowmelt runoff can cause severe spring flooding in the north-central and -eastern United States. Recent snowmelt floods in 1997, 2009, 2011, and 2019 flooded roads and washed-out bridges, resulting in large societal and economic impacts on communities throughout the US. Civil engineers and water resources managers rely on estimates of future precipitation extremes when making hydrologic estimates of design floods to size infrastructure (e.g., bridges and other hydraulic control structures; Figure 6.9). For snow conditions, higher-resolution RCM simulations are necessary to account for the effect of topographic features on precipitation and temperature, in order to improve estimates of total precipitation and its partitioning into rainfall and snow, and simulate increased precipitation downwind of large inland water bodies such as the Great Lakes.

7

Uncertainty in Future Projections, and Approaches for Representing Uncertainty

Future projections are uncertain, for multiple reasons. Limits to scientific understanding of natural variability, structural and parametric uncertainty in scientific modeling, climate sensitivity, bias correction, and downscaling all play a role. The uncertainty due to human choices that will determine emissions of heat-trapping gases becomes increasingly important over time, to the point where it dominates the uncertainty in many aspects of global and regional change by the end of century. Quantifying how a given system will respond to a changing climate adds yet another layer of uncertainty that can be prohibitively large in systems that are complex and/or not well understood. Understanding the source of these uncertainties and how they can be addressed when applying downscaled climate projections to assess future impacts is essential to quantifying the range of future change and resulting impacts on human and natural systems.

7.1 Identifying the Need for Quantitative Future Projections

One of the most important ways climate change affects both human and natural systems is by exacerbating or intensifying the effects of preexisting stressors, risks, and challenges. It is, as the US. 2014 Quadrennial Defense Review (Perry and Abizaid 2014) called it, a "threat multiplier." Climate change amplifies resource challenges, including: the demand for and supply of water and energy; food security, including crop yields, agricultural productivity, pest risks and even the nutritional content of food; human health, through the direct effects of heat as well as the indirect effects on air quality and weather and climate extremes; ecosystem health and services, including the spread of invasive species and the risk of extinctions; and more.

For this reason, quantifying the risks climate change poses to a given location, system, or sector begins with assessing the challenges and concerns that already exist today or may soon present themselves in the future. These can be varied, including: coping with the needs of a growing population; addressing socioeconomic disparity and social justice issues; improving air and water quality, and addressing other stressors on health or disease; maintaining the integrity of infrastructure and the built environment; and ensuring the continued health and function of critical ecosystems.

The next step is to connect these concerns directly to aspects of weather and climate that may affect them. These could include: shifts in the amount or timing of precipitation that affect water demand and supply; increases in the frequency or intensity of heat waves that exacerbate stress on energy infrastructure; more frequent heavy rain events, which could increase the risk of flood; or even a slow but steady increase in temperature that shifts the growing zones for crops or the geographic boundaries of key species or ecosystems.

The third step is to determine the nature of the climate information required to determine whether climate change is likely to significantly affect that system. In some cases, qualitative information, such as that described in Chapter 3, may be sufficient as the decisions that would make a given system more resilient in a changing climate are affected by the direction of change, but not the magnitude. In other cases, the type of quantitative climate projections discussed in this book may not be helpful, or necessary, or even appropriate because while accurately characterizing the uncertainty in quantitative climate projections is both important and challenging, sometimes this is not the most important aspect of the uncertainty. Here are three examples of situations where uncertainty in quantitative climate projections may not affect the outcome of the planning process.

First, a given system may be too complex or too poorly understood to be able to determine how changes in climate might affect it. In this case, high-resolution climate projections will not provide actionable information that decision makers could use to inform management or adaptation strategies. For example, while the assessments of climate impacts on urban regions have focused on sector-based analysis, urban system are a complex system of many interacting issues ranging from demographics, transportation and public health with climate change as one additional stressor (Ruth and Coelho 2007). In some cases, the greatest uncertainty in understanding the future trajectory of the system lies in understanding the dynamics of the system itself, including which factors (climatic and non-climatic) affect it.

Second, a system may be more sensitive to non-climate-related factors than climate factors. For example, return flows in a water district may be primarily influenced by population and use patterns and only slightly affected by climatic factors. In this case, future projections of return flows may be nearly identical under both historical and future climate conditions, and high-resolution climate projections would not add significant value to the project. Here, the greatest uncertainty lies in understanding the non-climatic drivers of the system, which can be related to human choices as well as natural factors, including population and demographics, socioeconomic factors, changes in land cover and land use, soil dynamics, and more.

Third, a system may be very sensitive to a climate- or weather-related risk, but that risk is not well quantified. For example, tornadoes can be devastating, and there is some indication that the so-called tornado alley in the central United States is shifting northeast and multi-tornado outbreaks are becoming more frequent; but linking observed trends to high-resolution modeling is a challenging task at the cutting edge of science today, and it is as yet unclear how the risk of tornadoes in a given location may be affected by a changing climate. In this case, the greatest uncertainty is scientific: the factors affecting the response of a given aspect of weather or climate to climate change over those regions are too complex and/or insufficiently well understood to predict the direction of future change with confidence. However, this uncertainty is so large that it precludes the development of meaningful and relevant quantitative projections.

In many cases, however, it is possible to generate quantitative climate projections that are able to constrain an envelope of projected future changes relevant to a given project, system, or region being studied. Additional discussion of the uncertainty in climate projections is provided by IPCC-AR5 (2013), Knutti and Sedláček (2013d), Hayhoe et al. (2017), and many of the national assessments discussed in Chapter 3. The remainder of this chapter discusses the sources of uncertainty in quantitative climate projections that contribute to this envelope of uncertainty: natural variability in weather and climate over timescales of hours to decades; scientific understanding of the climate system and how sensitive it is to rapidly increasing levels of carbon dioxide and other heat-trapping gases in the atmosphere; limitations in modeling climate and its response to both human and natural factors; and finally, the human choices that are the primary drivers of climate change today.

7.2 Uncertainty due to Natural Variability

Naturally occurring variability in the climate system is primarily the result of nonlinear processes and interactions within and between different components of the climate system, such as the exchange of heat between the atmosphere and ocean. This type of variability is referred to as "internal," since it occurs within the climate system, and "chaotic," as it is very sensitive to the initial conditions of the atmosphere–ocean system (Wigley and Raper 1990; Deser et al. 2012; Deser et al. 2014;). However, chaotic does not mean entirely unpredictable. Natural variability is also responsible for cyclical patterns or modes of variability that shift back and forth between positive and negative phases over timescales ranging from a few years to multiple decades.

One of the best-known natural modes of variability is the El Niño-Southern Oscillation, a recurring pattern of warmer and cooler sea-surface temperatures that extends westward across the Pacific from the coast of Peru. During the warm or positive phase, referred to as El Niño, sea-surface temperatures in that region are warmer than average, causing a net transfer of heat from the ocean to the atmosphere. As a result, global air temperatures are typically warmer than average during an El Niño year. El Niño also affects atmospheric circulation patterns, bringing drier-than-average conditions to some parts of the world, such as Australasia and India, and wetter-than-average conditions to other parts of the world, such as the contiguous United States during winter and spring (Figure 7.1). During the cool or negative phase, referred to as La Niña, global temperatures are typically cooler than average. Southeast Asia tends to be wetter than average, and the southwest and central United States tends to be drier than average. Other modes of variability that also affect regional temperatures and precipitation over timescales ranging from years to decades include the Pacific

El Niño and Rainfall

El Niño conditions in the tropical Pacific are known to shift rainfall patterns in many different parts of the world. Although they vary somewhat from one El Niño to the next, the strongest shifts remain fairly consistent in the regions and seasons shown on the map below.

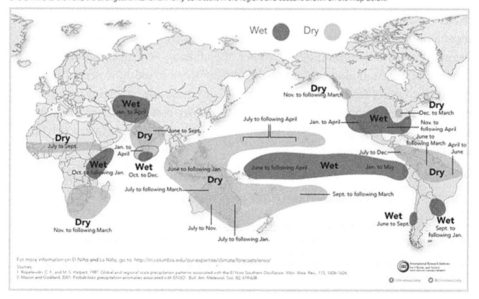

Figure 7.1 During an El Niño event, the positive phase of the ENSO cycle, rainfall patterns across the world shift, with some areas becoming wetter, and others drier. NOAA Climate.gov www.climate.gov/news-features/blogs/enso/how-enso-leads-cascade-global-impacts

Decadal Oscillation, the North Atlantic Oscillation, and the Atlantic Multi-decadal Oscillation.

Uncertainty in climate projections due to internal variability is most important over shorter time scales (less than twenty to thirty years) and finer spatial scales (Hawkins and Sutton 2009). This type of uncertainty is why climate projections are typically considered over longer time frames: by using several decades' worth of information, it is possible to average over multiple cycles of many of the most important modes of natural variability. Over these time scales, the impact of long-term climate trends become more prominent.

It is important to note that the more "extreme" or rare the climate event, the more important the role of natural variability in understanding the uncertainty in future projections. Incorporating not only a climatological period but also multiple GCM simulations into future projections provides a larger sample size from which to evaluate projected changes in the frequency and/or intensity of a given event. For example, say that a given piece of infrastructure was being designed to withstand the one-in-a hundred-year precipitation event. For historical conditions, such an extreme threshold is typically derived from a theoretical distribution fitted to observational data, since many records are not even a century long. However, by using multiple climate simulations, each with their own pattern of natural variability – say, twenty simulations for a twenty-year period – a sample size of 400 years is created, from which the one-in-a-hundred-year event can be empirically calculated or, if a theoretical distribution is fitted to the projections, will be built on a more robust sample.

7.3 Scientific Uncertainty

Scientific uncertainty is the result of scientists' evolving ability to understand and predict the response of the carbon cycle and climate system to global change. This uncertainty can be divided into three parts: climate sensitivity, structural uncertainty, and parametric uncertainty.

7.3.1 Climate Sensitivity

The first source of scientific uncertainty relates to the question of how sensitive the climate system is to the inadvertent yet nonetheless unprecedented experiment humanity is conducting with it. The direct effect of increasing carbon dioxide levels in the atmosphere can be calculated relatively precisely (Myhre et al. 1998). The net effect, referred to as climate sensitivity (Schlesinger and Mitchell 1985; Wilson and Mitchell 1987), however, depends on the feedbacks or self-reinforcing cycles that can amplify, or in some cases diminish, the initial warming as the climate system responds over time scales ranging from months to millennia. Processes that occur over time scales relevant to human decision making include changes in the reflectivity or albedo of the Arctic due to shrinking sea-ice extent, increasing emissions of carbon and methane from thawing permafrost, and changes in the physical properties of important elements of the climate system, from cloud characteristics to the inflow of freshwater to the ocean, which affects ocean circulation and, in turn, the transport of heat around the world.

It is impossible to determine the exact value of climate sensitivity as it depends on the initial state of the climate system as well as on the magnitude and type of forcing – today, emissions of heat-trapping gases from fossil-fuel use and land use including agriculture and deforestation. In terms of the magnitude of forcing, past observations and paleoclimate data show no precise analog to the present-day conditions. While at other times in the earth's history, the planet has been as warm or warmer than today, at no time has this much carbon been transferred directly into the atmosphere this rapidly. The closest analog is the Paleocene–Eocene Thermal Maximum (PETM), approximately 55–56 million years ago. However, it is estimated that the maximum annual rate of sustained carbon release into the atmosphere at that time was around 10 percent of human carbon emissions today (Crowley 1990; Zeebe et al. 2016). However, combining paleoclimate data, observations, models, and physical under-standing of the climate system, scientists can estimate a range of values for equilibrium climate sensitivity (ECS). ECS is formally defined as the increase in global near-surface air temperature when carbon dioxide doubles relative to preindustrial times. According to the IPCC, there is "medium confidence that the ECS is likely between 1.5 °C and 4.5 °C," and this range has been remarkably consistent since the first initial estimates of ECS, made decades ago, based on less data and much simpler GCMs (e.g., Schlesinger and Mitchell 1985).

It is also important to note that ECS does not encompass the full response of the climate system to human-induced warming. Earth System Sensitivity (ESS) factors in the slower interactions between various components of the earth's climate system, such as the responses of the deep ocean and the Greenland and Antarctic ice sheets to warming that occur over centuries to millennia (Figure 7.2). It is estimated that ESS is likely approximately double the value of ECS, and, as Kopp et al. (2017) caution, "while climate models incorporate important climate processes that can be well quantified, they do not include all of the processes that can contribute to feedbacks, compound extreme events, and abrupt and/or irreversible changes. For this reason, future changes outside the range projected by climate models cannot be ruled out. Moreover, the systematic tendency of climate models to underestimate temperature change during warm paleoclimates suggests that climate models are more likely to underestimate than to overestimate the amount of long-term future change."

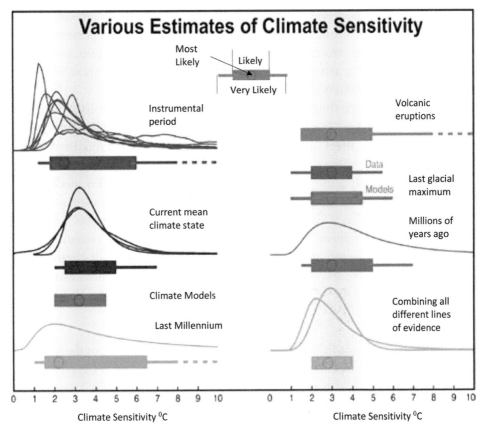

Figure 7.2 The probability distribution of ECS as estimated from various sources of data and information.
Source: Knutti and Hegerl (2008)

7.3.2 *Structural Uncertainty*

The second source of scientific uncertainty relates to the question of the extent to which scientists are able to understand, quantify, and accurately represent the components of the climate system and the interactions between them in GCMs. These components and the processes that connect them are what determine climate in every location around the globe and how it responds to both natural and human-caused changes. Because scientific understanding of the climate system is still evolving, certain aspects of the Earth system may not be accurately or completely represented, or an important process might not be included. This could be because that process is not yet known, or because it is known, but is not understood well enough to be modeled accurately. For example, it is likely that there are processes as yet unknown that have led to the very rapid climate changes documented in the paleoclimate record. On the other hand, although the existence of important dynamical mechanisms driving ice-sheet melt is known, GCMs do not yet include a full description of them. Currently process-based models representing the dynamics of ice sheet are in development and being tested in climate models (Lispcomb et al. 2013), and a comparison of the ice-sheet models (Nowicki et al. 2016) from several CMIP6 ESMs is ongoing.

As scientific understanding and computing power increases over time, more and more physical processes are incorporated into GCMs (see Chapter 2 for more discussion). This can affect the range of climate sensitivity of the models: CMIP6 models, for example, show a higher sensitivity to increasing carbon levels in the atmosphere than CMIP5 models (Eyring et al. 2019). More CMIP6 models also include an interactive carbon cycle; many have also significantly increased their spatial resolution, which affects the representation of physical processes; and some have altered their treatment of sensitive aspects of the climate system such as clouds. All of these factors make up the structure of the model and contribute to uncertainty in assessing the response of the climate system to future change.

7.3.3 Parametric Uncertainty

The third source of scientific uncertainty relates directly to the issues of the temporal and spatial scale at which GCMs and RCMs operate. As discussed in Chapters 2 and 4, processes that occur at timescales or spatial scales below the resolution of the model must be parameterized (McFarlane 2011). Sub-grid scale processes typically include cloud formation, precipitation, turbulence, and the effects of dust on the atmosphere. The results of lab experiments, ground-based observations, observations from aircraft-based platforms to study atmospheric processes inside clouds and aerosols, and high-resolution modeling are used to understand how these processes appear in aggregate, at the scale of a GCM or RCM and contribute to the development of sub-models that parameterize processes occurring at scales that are smaller than those at which observations are made. As parameterization by definition results in an incomplete and imperfect representation of these smaller-scale processes, the sub-model may not hold true for every location or under all possible conditions, particularly as climate continues to change in the future. The use of these parameterizations to represent physical processes at as yet unresolved scales in the models is known as parametric uncertainty (Sanderson 2011).

7.3.4 Accounting for Scientific Uncertainty

Scientific uncertainty, as the net result of uncertainty in climate sensitivity, model structure, and parametrization of sub-grid-scale processes, is important in determining the magnitude and sometimes even the direction of projected changes in many variables, particularly at the local to regional scale, including seasonal average precipitation and the frequency and intensity of many types of extremes. Climate sensitivity is important at the global scale and also broadly at the regional scale, although it is well understood how and why land areas are warming faster than oceans and higher-latitude regions faster than the tropics. Both structural and parametric uncertainties tend to be more important for predicting changes at the regional as compared to the global scale over the rest of the century. (Over longer time scales, as discussed in the quote from Kopp et al. (2017), structural uncertainties may be significant at the global scale as well.) They are also usually – but, as discussed in Chapter 4, not always – more important when simulating aspects of climate related to precipitation (where small-scale cloud processes are very important) as compared to temperature, which is affected primarily by large-scale dynamics that can be resolved by the models.

To account for scientific uncertainty in assessing future impacts using quantitative, high-resolution climate projections, simulations from as many GCMs as possible should be used (Hayhoe et al. 2017). First, it is important to consider the range of climate sensitivity covered by the GCMs. Using GCMs with different climate sensitivities ensures the simulations encompass a reasonable range of scientific uncertainty, from higher to lower. Second, additional differences between the models represent the limitations of scientific ability to simulate the climate system, so for any given scenario, a more robust range of likely outcomes will be obtained from a larger set or *ensemble* of GCMs. The default for any analysis with sufficient resources is to consider simulations from all GCMs with appropriate outputs. In reality, however, practical considerations may limit the number of GCMs and/or simulations that can be used, including which GCMs are available from the database of downscaled or high-resolution projections. If the number of GCMs must be constrained by the limits of the project, then it makes sense to favor the most recent versions of reliable, well documented, long-established and well-tested GCMs from modeling groups with decades of experience that are in their third, fourth, or even fifth generation. Less priority can be placed on GCMs that are entirely new or have not yet been extensively evaluated in the peer-reviewed literature, and/or originate from modeling groups with less experience.

Sometimes, analyses attempt to identify a single "best" or set of "better" models for a certain region or variable. However, such attempts are challenging, can yield misleading results, and do not guarantee that a model or set of models will be better able to simulate projected changes in those same variables or for the same region than a larger multi-model ensemble. The GCM weighting schemes used in NCA4 and upcoming IPCC AR6 (see Chapter 2) include metrics that evaluate the basic abilities of models but also focus on the extent to which GCMs share aspects of their architecture or design. Applying these weighting schemes when averaging across a large ensemble of GCMs to create a multi-model average provides more robust results, primarily by identifying the GCMs that cannot be treated as entirely independent models (i.e. given a weight of one).

7.4 Uncertainty due to Human Choices

Scenario uncertainty is the result of the fact that human activities are the primary driver of climate change today and for the foreseeable future. The future emissions of carbon dioxide and other heat-trapping gases from human activities that are driving this change depend on demographics, economic and technological development, energy use, and policy choices. Most of these are difficult, if not impossible, to predict with any certainty far into the future. Instead of attempting to predict human behavior, climate projections rely on multiple scenarios that capture a broad range of possible outcomes: from a fossil-fuel-intensive future where carbon emissions continue to grow, to one where emissions are reduced rapidly and become net negative (i.e. carbon uptake from human activities exceeds emissions) before the end of the century.

7.4.1 Scenarios Used for GCM Simulations

Box 7.1
Adapting to Sea Level Rise

The South Florida community is undertaking a holistic approach to adaptation planning using regional sea level rise projections and hydrologic modeling to inform integrated approaches to land use, resource management, and infrastructure planning. Climate information is obtained by comparing statistical and dynamically downscaled climate projections [CMIP5, LOCA (Localized Constructed Analogues; see Chapter 5), COAPS (Center for Ocean-Atmospheric Prediction Studies, Florida State University, Tallahassee, Florida, USA) and CORDEX (see Chapter 5)] datasets to inform the development of rainfall intensity-duration-frequency curves, annual maximum seasonality and trends, and the temporal distribution of heavy precipitation. If bias remains large (as has been the case with GCM output), the County plans to employ a bracketed, scenario-based approach. The results will inform a regional risk assessment and resilient infrastructure improvement plan with basin level infrastructure improvements and planning level cost estimates. The City of Fort Lauderdale is already undertaking a similar stormwater improvement plan based on regional model results using the dynamically downscaled COAPS data set as input for high-resolution hydrologic modeling.

The previous-generation CMIP3 simulations that provided climate projections used in the Third and Fourth IPCC Assessment Reports and the Second and Third US National Climate Assessments, and many other published impact studies, are based on the emission scenarios from the Special Report on Emission Scenarios (SRES; Nakicenovic et al. 2000). The main SRES scenarios consist of the higher A1FI (FI stands for fossil-intensive), mid-high A2 and A1B, mid-low B2 and A1T, and lower B1. Projected changes in global temperature under the SRES scenarios as simulated by CMIP3 GCMs range from about 2 °C under a lower scenario to more than 5 °C under a higher scenario by end of century. None of the SRES scenarios represent a future with policies specifically designed to reduce emissions of carbon dioxide or other human emissions.

The CMIP5 simulations that provided the climate projections used in the Fifth IPCC Assessment Report and the Third and Fourth US National Climate Assessments are based on Representative Concentration Pathways (RCPs; Moss et al. 2010). The main RCP scenarios used in CMIP5 consisted of the higher RCP8.5, mid-high RCP6.0, lower RCP4.5 and even lower RCP2.6. The number after RCP stands for the amount of radiative forcing from human emissions by end of century, in units of watts per square meter; a higher value means greater forcing and greater subsequent change. At the higher end of the range, RCP8.5 is very similar to SRES A1FI; at the lower end of the range, RCP4.5 is very similar to SRES B1. However, all of the three lower RCP scenarios (2.6, 4.5, and 6.0) are climate-policy scenarios and RCP2.6 specifically envisions a future where successful policies to deliberately and rapidly reduce carbon emissions bring net carbon emissions to zero around 2070, with net negative emissions (i.e. carbon uptake) after that. The carbon emissions and global temperature corresponding to each main RCP scenario is shown in Figure 7.3.

The CMIP6 simulations that provide the climate projections used in the Sixth IPCC Assessment Report are based on RCP scenarios combined with Shared Socioeconomic Pathways (SSPs). SSPs provide projections of population, demographics, economics, and technology consistent with a given RCP to support the analysis of impacts and vulnerability at the global to regional scale (O'Neill et al. 2014). The five SSPs are: 1, sustainability; 2, middle of the road; 3, regional rivalries; 4, inequality; and 5, fossil-fueled development. For CMIP6, these SSPs were combined with the preexisting RCPs to create four main scenarios: SSP1 with RCP2.6, referred to as SSP126; SSP2 with RCP4.5, called SSP245; SSP3 with RCP7.0 (replacing the former RCP6.0 with one where radiative forcing by end of century reaches 7.0 watts per square meter instead of 6.0), called SSP370; and SSP5 with RCP8.5, called SSP585.

While the different families of scenarios and their labeling or numbering can seem confusing and even contradictory, in reality these approaches are not

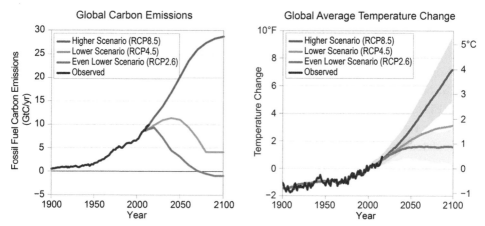

Figure 7.3 Observed and projected changes in global average temperature (right) depend on observed and projected emissions of carbon dioxide from fossil-fuel combustion (left) and emissions of carbon dioxide and other heat-trapping gases from other human activities, including land use and land-use change. This figure shows the carbon emissions and associated change in global average temperature under the higher RCP8.5, lower RCP4.5 and even lower RCP2.6 scenarios.
Source: Hayhoe et al. (2017)

inconsistent. Rather, they all present a wide range of plausible and consistent pictures of how future human activities may affect climate. Projected impacts under a higher future, where humans continue to depend on fossil fuels, show the greater amount of change that may be expected, and allow planners and decision makers to demonstrate the benefits of mitigation in terms of impacts avoided. Projected impacts under a lower future, where carbon emissions are significantly reduced, allow planners and decision makers to characterize the minimum change that will be expected and furthermore quantify the benefits of reducing emissions.

7.4.2 Addressing Scenario Uncertainty in Impact Assessments

An additional 0.6 °C of warming beyond what has already been experienced is nearly inevitable, because of past emissions and current energy infrastructure (IPCC 2013; Hayhoe et al. 2017). Because of this, the proportion of the uncertainty in climate change experienced at the local to regional scale over the next twenty years or so that is specifically due to scenarios is minimal. This is not to say there will be no impacts of any kind from human choices over this time; the shifts in population, energy production and consumption, and land use encapsulated by the SSPs can have immediate effects on air quality, water demand, ecosystem health, local climate, and the local and regional economy.

However, the impact of human choices today and over the next few decades on global and regional climate will not be immediately obvious, for two reasons. The former is the technological and economic inertia of energy infrastructure to policy changes. For example, policies or incentives may be put in place to transition from gas-powered to electric cars, or from coal to wind. However, it takes some time for the transition to be fully realized, as ceasing the use of vehicles or power plants that are still economically viable is typically cost-prohibitive. The latter reason is the inertia in the climate system response to human emissions. As discussed previously in Section 7.3, the direct effect of carbon emissions on warming is only part of the total warming. It takes time for the climate system to fully respond to an increase of heat-trapping gases in the atmosphere: years to decades for some aspects of the system, such as Arctic sea-ice or cloud properties, and centuries to millennia for others, such as the heat content of the deep ocean or the melting of the Greenland or Antarctic ice sheets. Thus, the impact of policies and associated emission reductions made today on global climate will not be evident for several decades.

A substantial amount of future change, however, may be avoided by significantly reducing and eventually eliminating carbon emissions. Over longer timescales, beyond thirty years or so, the importance of scenario uncertainty, and of the choices, made today and in the near future, becomes increasingly evident in climate projections. This provides the motivation for international efforts such as the Paris Agreement, which aims to hold "the increase in the global average temperature to well below 2 °C above preindustrial levels and to pursue efforts to limit the temperature increase to 1.5 °C above preindustrial levels, recognizing that this would significantly reduce the risks and impacts of climate change."

Quantifying this uncertainty by identifying a "more likely" scenario, however, is even more problematic than attempts to identify a "best" GCM, as it is impossible to predict human behavior with certainty. At the upper end of the range, even though RCP8.5 and SSP585 reflect the forcing associated with the upper range of published emissions estimates, this scenario is not intended to serve as an upper limit on possible emissions nor as a business as usual or reference scenario for the other scenarios, and the IPCC specifically states that no probabilities should be inferred from the distribution of the scenarios. In other words, a scenario in the middle of the range is not any more likely than a scenario at the edge of the range. At the lower end of the range, RCP2.6/SSP126 is broadly consistent with the listed goals of the Paris Agreement listed. However, comparing observed carbon emissions to those projected to occur under the RCP scenarios since 2006, Hayhoe et al. (2017) conclude that, "the observed increase in global carbon emissions over the past 15–20 years has been consistent with higher scenarios (very high confidence). In 2014 and 2015, emission growth rates slowed as economic growth has become less carbon-intensive (medium confidence). Even if this trend continues, however, it is not yet at a rate that would limit the increase in the global average temperature to well below 2°C above preindustrial levels (high confidence)." And it is possible that the actual trajectory, such as a continued rise in carbon emissions until the impacts become so obvious and urgent that policies are implemented that cause an abrupt decrease, may follow none of the scenarios.

To account for scenario uncertainty in most projects and applications, at least two future scenarios should be used to cover a range of possible outcomes. Some questions can be answered with only a lower scenario (e.g., what is the minimum amount of increased flood risk a city will have to adapt to, even if carbon emissions are rapidly reduced?), or a higher scenario (e.g., what are the consequences for coastal sea level rise of continuing a fossil-fuel-intensive pathway?). Most assessments, however, encompass the range of possible futures by quantifying expected impacts under a higher scenario where emissions continue to grow, consistent with trends over the last few decades, and impacts under a lower scenario where emissions are sharply reduced, consistent with policies or targets such as the Paris Agreement.

An alternative approach to quantifying impacts under a certain scenario (e.g., RCP8.5/SSP585 and RCP4.5/SSP245) for specific future climatological time periods (e.g., 2030–49, 2080–99) is to specify impacts for a given global mean temperature threshold (e.g., +1.5, +2, +3 °C above preindustrial levels) instead. Although these are not directly linked to a specific time frame, these thresholds will be reached more quickly if human emissions continue to grow, and more slowly (or not at all) if emissions are sharply reduced. The magnitude and severity of many impacts depends more on the amount of global change than on the rate at which it occurs. Physical changes at the regional scale that have been shown to scale with global mean temperature include average temperature, extreme heat, runoff, drought risk, area burned by wildfire, temperature-related crop yield changes, and more (NAS 2012; IPCC 2013). As described in Hayhoe et al. (2017), GMT scenarios can be calculated from GCM simulations based on SRES, RCP, or SSP-RCP scenarios, and offer an alternative approach to quantifying the projected impacts that could occur under a given amount of warming, regardless of when that may occur. Furthermore, by quantifying projected changes for a given amount of warming regardless of when it may be reached, this approach de-emphasizes the uncertainty due to the sensitivity of the global climate system to those emissions and connects adaptation, resilience, and vulnerability assessments directly to global targets such as the Paris Agreement, which are expressed in terms of global mean temperature.

7.5 The Relative Importance of Different Sources of Uncertainty

Natural variability, scientific uncertainty, and scenarios or human uncertainty represent three important sources of uncertainty in climate projections at the global to regional scale that can be relatively more or less important for a given application, depending on the time scale, the spatial scale, and the variable being considered. In terms of time frame, uncertainty due to the internal variability of the climate system is more important over the short term (up to fifteen years for temperature and forty years for precipitation, depending on location) and at smaller spatial scales. Scientific uncertainty is more important on longer timescales (beyond fifteen to forty years) and for changes such as ice-sheet melt and the rate of sea level rise that depend critically on processes that are not yet well quantified. Scenario uncertainty becomes increasingly important after mid-century, and for average temperature and many types of extremes, including extreme heat and heavy precipitation, as compared to average precipitation. In general, the range of uncertainty will be larger for projected changes in extremes and rare events as compared to in mean values, and for precipitation as compared to temperature.

Quantifying the relative contribution of each of these sources of uncertainty – natural variability, scientific, and scenarios – to overall uncertainty in future climate projections illustrates the importance of time scale and geographic location in identifying the most relevant sources of uncertainty in future projections (Figure 7.4). For temperature, it is clear that increasing emissions from human activities will drive consistent increases in global and most regional temperatures. Although natural variability may affect the rate of global temperature change over timescales up to a decade, uncertainty in long-term temperature trends is primarily a function of future emissions from human activities, and the sensitivity of the Earth's climate system to those emissions. Scientific uncertainty is more important for mid-century temperature projections – up to forty years in the tropics and mid-latitudes, and up to eighty years for higher latitudes – and remains critical for annual precipitation through the end of the century.

It is important to note that uncertainty arises in the translation of the impacts of global change down to the regional scale. At smaller spatial scales, the influence of natural variability tends to be greater. There are also other natural and human factors that can complicate the relationship between local climate and global change, ranging from the impact of natural cycles such as El Niño on regional precipitation to the influence of urbanization or land-use change on local climate. And as discussed in Chapters 4 and 5, each type of downscaling has associated uncertainty and error, and as discussed in Chapter 6, determining the added value of high-resolution projections depends on understanding how these compare to the confidence that can be placed in the direction and magnitude of that change.

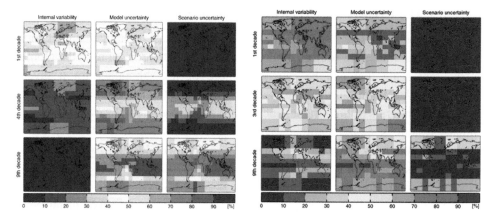

Figure 7.4 These maps show how the dominant sources of uncertainty in decadal mean annual surface temperature (left) and precipitation (right) change over time, the further out into the future the projections extend. Even on regional scales, the uncertainty due to internal natural variability (left column of each plot) is only a significant component for lead times up to a decade or two for temperature and three decades for precipitation. By the end of the century, scenario uncertainty is the most important factor in determining temperature, except at high latitudes where self-reinforcing feedbacks are particularly important, and scientific uncertainty continues to be the most important factor in determining changes in annual precipitation, particularly in the tropics.
Source: Hawkins and Sutton (2009; 2011)

7.6 The Importance of Quantifying Uncertainty

To be useful, climate projections and their associated uncertainty should enable planners, engineers, managers or other decision makers to weigh the consequences of their choices more accurately than they are able to without this information. Overestimating uncertainty can lead to unrealistic projections that could be used to recommend potentially unnecessary expenditures or as a justification for failure to act. However, it may be an even greater risk to underestimate the uncertainty, and to consequently be unprepared for or unable to mitigate the impacts of low-probability, high-cost events. Very rare events made up of multiple reinforcing factors, such as a combination of record-setting drought, high temperatures, and wildfire, can have potentially devastating impacts (Kopp et al. 2017).

The first type of regret is over-preparation, or what is known as a Type I error: over-investment in the near term for an impact that was predicted but does not occur or, if it does, is much less than predicted. The second type of regret is under-preparation, or what is known as a Type II error: little or no investment in the near term, with potentially expensive or even devastating consequences.

How might these errors occur? One example is scenario selection. Projections based on a lower or mid-range scenario run the risk of under-preparation, if emissions of heat-trapping gases continue on their current trajectory rather than being sharply reduced, or even if emissions decrease but climate sensitivity, the response of the climate system to a given amount of human-caused warming, is greater than expected. Projections based only on a higher scenario toward end of century run the risk of over-preparation, if nations are able to meet the goals of the Paris Agreement and the climate system response is in the middle to lower end of the expected range.

Another example is model selection. Projects run the risk of either of these types of errors if they select projections based only on one or two GCMs, particularly if the GCMs selected tend to have climate sensitivity values that are on the higher or lower end of the range of likely values. If this is the case, the resulting projections would be, on average, at the lower or higher end of the range of projections that would result from a larger multi-model ensemble. (It is important to note, however, that GCMs do not cover the full range of climate sensitivity and, particularly over longer periods of time, are more likely to under- than over-estimate the magnitude of future change; Kopp et al. 2017).

Third, even if it were somehow possible to accurately predict human emissions and the response of the climate system to those emissions, it is still impossible to completely eliminate the risk of either over- or under-estimating future impacts. This is because of the chaotic natural variability in the climate system. In general, the likelihood of error decreases as longer time periods and/or greater numbers of GCM simulations are used in an analysis; quantifying changes for climatological periods of at least twenty or thirty years and using a large ensemble of simulations increases the skill, reliability, and consistency of GCM-based projections (Tebaldi and Knutti 2007). Ensuring a sufficiently large sample of natural variability is particularly important for relatively rare extremes, where relying on too few data points could seriously misrepresent the likelihood of extremes in either direction.

While it is impossible to entirely eliminate either type of error, understanding the nature of the project or system being studied can also help inform the importance of uncertainty. For example, some resilience strategies, adaptation actions, or systems may be relatively insensitive to over-preparation. When installing cooling centers to help an urban population adapt to increasing risk of more severe heatwaves, even if heatwaves were predicted to be three to five times more frequent over the next few decades but it turned out that the heatwaves were only two to three times more frequent, the solution would still have been nearly identical and there would be few regrets. Another system may be relatively insensitive to under-preparation; if adaptation to sea level rise in a coastal area where most infrastructure lies significantly above sea level were made today, vs. twenty years from now, there would be very little difference and little regret. On the other hand, some projects may be very sensitive to these types of errors;

installing a new storm-sewer system, for example, is extremely costly. If heavy precipitation in the region is increasing, but far larger pipes are installed than end up being needed over the lifetime of the system, too much money will have been spent up front. On the other hand, if too small pipes are installed and the entire system has to be replaced prematurely, the costs would be even greater. In such a case, a substantial investment in a large ensemble of detailed, quantitative projections, possibly even coupled with additional analysis of the physical processes driving changes in heavy precipitation over that region, could be worthwhile.

In conclusion, there are many sources of uncertainty in quantifying future impacts from climate change at the local to regional scale. Understanding the sensitivity of the system to climate, and how the relative importance of the various sources of scientific and scenario uncertainty vary over time and by geographic location is key to ensuring they are accounted for appropriately in a given project or analysis.

8

Guidance and Recommendations for Use of (Downscaled) Climate Information

Previous chapters have laid out the process of creating high-resolution climate projections at spatial and time scales appropriate for assessing the impacts of climate change at the local to regional scale, and described available assessments, global climate model archives, and both RCM- and ESDM-based high-resolution projections. This chapter builds on this information to discuss the process by which high-resolution climate projections can be selected, applied, and interpreted to quantify future impacts for a given location, system, or assessment.

8.1 Introduction

Many people, picking up this book, may have been hoping for a clear and brief guide to the "best" information. They might have even preferred a booklet or a decision tree. But if you have made it this far, you know that there is no such universal conclusion and it is impossible to summarize all the information that goes into making these choices in a few short pages or an automated tool. You also know that it is not impossible because we do not know how, but rather because we *cannot condense this to a simple recipe.* And in some cases, the methodology used to make decisions must be changed or adapted to non-stationarity: 5-minute wind speeds are not available through end of century.

Since the goals and resources for one application may be very different from those of another, the most appropriate input for a given study, application, or purpose depends on multiple factors. What would be considered the "best" or most appropriate method depends on the user's needs, abilities and resources. It can vary according to the spatial and temporal resolution of the desired information and the climate variables that have the greatest effect on the region and/or sector being considered. It also depends on the question being asked, and the resources the user has. Factors to consider include:

- *Spatial Resolution.* An analysis for a region, state, or province might use a gridded dataset that covers the whole region with a resolution of several kilometers or miles per grid cell; an analysis for a city or a lake or reservoir might use data from the nearest weather station(s) as these best represent local conditions over a smaller area.
- *Temporal Resolution.* Hydraulic engineers examining extreme precipitation and its impact on flood risk in an urban area might require precipitation data at the hourly scale, or even finer; ecologists assessing the impact of drought on a tropical ecosystem in a given region might only require precipitation data for the dry and wet seasons.
- *Relevant Climate Variables.* Climate inputs to an agricultural yield model may include maximum and minimum temperature, precipitation, humidity, solar radiation, and wind. An envelope model that determines the geographic range of an endangered or invasive species that is limited by cold or warm temperatures may only require maximum or minimum temperature.
- *Question Asked.* Two different organizations might require information on how water supply and demand may vary in the future; but the first organization is only planning for the next ten years, with a focus on year-to-year variability, whereas the second is planning for the next fifty years, with a focus on long-term trends toward drier or wetter conditions. For the first organization, recent historical data on observed trends and variability, coupled with uncertainty estimates due to natural variability obtained from multiple climate model simulations may be useful. For the second organization, it will be essential to use simulations based on multiple future pathways, to quantify the long-term impacts on water supply if the world continues to depend on fossil fuels vs. impacts if the world transitions to clean energy.
- *Technical Capacity and/or Resources.* In assessing the impacts of climate change on the urban heat island and public health, one organization might have a limited time and budget, such that accessing publicly available information on how many days per year temperatures would pass an extreme temperature threshold (e.g., 100 °F or 40 °C) and combining this with information on how our health is affected by heat and what we can do to prepare and adapt to it would be sufficient. Another organization might want to create heat maps by block for the entire city in order to inform infrastructure and urban-planning decisions; this project could require expertise in data and analysis programs including ArcGIS, the netCDF files typically used to store climate data, statistics used to spatially interpolate data, and access to block-level infrastructure and demographic information.
- *Limitations to Available Resources.* Data availability changes dramatically from one location to another. A very densely populated region such as eastern Massachusetts or southern Ontario will have much more historical data available than a

region such as Big Bend National Park or an entire province such as Nunavut, where there may not be a long-term weather station for over a hundred miles or hundreds of kilometers. Even if gridded data are available, they will be less reliable due to the lack of observational data to ground-truth it. In terms of future projections, many datasets offering both statistically and dynamically downscaled projections are available for North America and Europe, but very few if any are available for regions and countries in southeast Asia, Africa, and South America.

Some studies only seek to understand the sensitivity of a given system or region to a range of plausible future changes in mean climate. For these, published climate projections (for example, those used in the Fourth US National Climate Assessments (USGCRP 2017; 2018b) and in the atlas of climate projections produced by the IPCC Working Group I (van Oldenborgh et al. 2013 – and planned for the next IPCC WG1 Report for IPCC Assessment 6) or even historical trends can inform adequate estimates of projected changes in regional temperature, precipitation, or sea level. Other studies seek to quantify the projected impacts of climate change for a given time frame or range of future scenarios. Many of these studies require quantitative climate projections as inputs, raising questions about method, model, and scenario selection.

We authors come from both the climate science and the user sides of developing relevant information for assessments and decision support. We have worked, together and separately, on dozens of such studies that have attempted – often successfully, but sometimes not – to quantify the impacts of climate change on a broad range of regions and sectors. We have looked at climate impacts on transportation infrastructure, military installations, winter tourism, city planning, agriculture, communications sector, energy use, endangered and invasive species, human health, and more. This experience has repeatedly reinforced the need for a specialized group of practitioners beyond the traditional climate scientists, who are typically focused on understanding the physical processes at work in the atmosphere, ocean, cryosphere, and biosphere today, to include scientists and experts specifically trained at the interface of climate science, data, risk, and adaptation science. One of our primary recommendations to anyone interested in using high-resolution projections of future climate, and lacking this internal expertise, is to work with applied scientists familiar with the various methods available and your region of interest. It is also important that this individual understands the needs of stakeholders. While we do provide guidance in the next sections, this is only in the form of general guidelines and cannot substitute for working with experts in the selection and use of future climate information.

This need for individuals who are both trained in the use of climate model output and who are familiar with a number of impact areas and the complexity of decision-making contexts has been documented (NAS 2012). These "climate interpreters" would act at the interface between climate researchers and the wide variety of users of climate information. Specifically, the NAS (2012) recommended the development of degree or certification programs in "climate interpretation." The role of interpreters would be similar to that of the "information broker," called for by Dilling and Lemos (2011). At the time of publication of this book, however, we are unaware of any such training or certification program or even any database of names to which to refer the user. Instead, in seeking such expertise, we currently recommend: (1) contacting others who have conducted or published similar assessments in your region, even if on a different sector, or for your sector, even if in a different region, to ask for their insight and recommendation (for North America, the Climate Adaptation and Knowledge Exchange has many such examples, and many more are published in the scientific literature or in reports and white papers available online); (2) reaching out to a local or regional university or research institution for relevant expertise and/or recommendations; (3) contacting a national organization such as the US Global Change Research Program, Environment Canada, the Australian Commonwealth Scientific and Industrial Research Organization, the UK Meteorological Office's Hadley Centre, the French National Center for Scientific Research, the German Potsdam Institute, the Norwegian Meteorological Institute, the Japanese Tokyo Climate Center, etcetera; and last but not least; (4) drawing on this book and the additional resources it cites for generalized insight and guidelines that can be interpreted and applied to your specific situation.

8.2 Global Climate Model Selection

For use in impacts and adaptation planning, in general one wants to sample the full range of uncertainty present in the projections produced by the many global models. At this point there are a large number of GCMs. For example, in CMIP5, there were at least thirty-five different models. Sampling the range of uncertainty does not mean that all thirty-five CMIP5 models would need to be included. As discussed in Chapters 2 and 8, there are different ways of sampling these models, based on, for example, their performance in producing the current climate and in terms of the range of projections of climate change produced by the different climate models. Because these two factors will vary with the region being considered, the subset of models may also differ by region.

8.3 Emission Scenarios

As also discussed in Chapters 2 and 7, there are a variety of RCPs currently being used to drive the various global climate models. The four major RCPs that have been used are RCP 8.5, 6.0, 4.5, and 2.6. To most completely consider the uncertainty across future pathways, theoretically all four RCPs would be used. But this also depends upon how far into the future one is considering impacts and adaptation. As described in Chapter 2 (Figure 2.8), in the first half of the twenty-first century, the effect of the different RCPs is much less important than the effect of using different climate models. As one proceeds later into the twenty-first century, the RCPs become more important, and it would become important to sample both the pathways and models to completely cover the uncertainty envelope. However, if one's goal was to determine the extreme case of adaptation, then one could decide to use only the highest RCP for an adaptation study, (i.e., if one could adapt to the highest RCP, then adaptation to the lower RCPs would also be possible). RCP and GCM selection is typically made in combination with downscaling choices. In that case, many of these choices have already been made in major scientific downscaling programs particularly when it concerns dynamical downscaling (e.g., CORDEX) (see Table 8.2).

8.4 Natural Variability

The uncertainty due to natural variability of the climate system has been less explored than the uncertainties due to modeling of the climate system and that of the future trajectory of greenhouse gases. However, it has been recognized as an important factor in recent years (e.g., Deser et al. 2014, Kay et al. 2015 and see also Chapter 7). Many global climate modeling centers now produce multiple realizations of climate simulations using different initial conditions and make these available primarily to the scientific community. Multiple RCM experiments have now been performed using different realizations of one GCM (e.g., von Trentini et al. 2019). There is also a great deal of interest in understanding impacts and estimating climate change risks in the thirty-to-fifty-year time frame, time frames where the natural variability is an important component of uncertainty. However, these have been few studies (e.g., Prudhomme and Davies 2009) that have used natural variability for impacts and adaptation studies. However, with increasing availability of GCM simulation ensembles that are designed to quantify the natural variability of the models, the use of these for impacts and adaptation studies will likely become more common.

8.5 Selecting Downscaling Approaches

It is not possible to provide clear, detailed recommendations for different downscaling methods that are appropriate for all assessment and planning efforts.

As discussed in Chapter 6, different methods produce different high-resolution climate changes (e.g., Spak et al. 2007; Vavrus and Behnke 2014; Tang et al. 2016). New approaches, such as a "perfect model" framework, also discussed in Chapter 6, are advancing the state of the science in downscaling intercomparison and evaluation (Dixon et al. 2016). In general, however, a downscaling method should be selected that can credibly resolve spatial and temporal scales that are relevant to the type of analysis desired and that can produce salient and credible climate information for the problem being investigated (Cash et al. 2002; Lemos and Rood 2010).

In some cases, downscaling may not be required at all because the GCMs adequately represent the relevant climate phenomena. For example, regarding temperature in flat terrain, the simple interpolation of the GCM to the required scale may be adequate (Maraun and Widmann 2018). As discussed in earlier chapters, there are two fundamental approaches to downscaling: dynamical and statistical. Each has its advantages and limitations. An understanding of those advantages and limitations can aid in identifying the most appropriate method to be applied to a given problem.

One of the reasons that it is not possible to provide detailed recommendations for what method to use for which problem is that there remains a paucity of research comparing the different methods in detail in different regions for different variables and evaluating their relative credibility.

Box 8.1
Case Study. New York City Adaptation Planning

The New York City Office of Recovery and Resiliency has developed guidelines for maintaining climate resiliency (NYC Mayors Office of Recovery and Resiliency 2019). The guidelines provide "step-by-step instructions on how to supplement historic climate data with specific, regional forward looking climate change data in the design of City facilities." They are intended to apply to all City capital projects (except for coastal ones, for which a separate guidelines document has been prepared). The guidelines make recommendations for capital projects regarding increasing heat, precipitation, and sea level rise. Climate change projections are provided by the New York CIty Panel on Climate Change (NPCC). The projections are based on thirty-five GCMs and two different representative concentration pathways (RCP 4.5 and 8.5). Values for the low end (10th percentile) middle range, and high end (90th percentile) are provided. The downscaling applied is the simple "delta" approach (Mearns et al. 2001) whereby the change in temperature for example, for the GCM grid box relevant to the subregion is appended to the higher resolution observed dataset used – see Figure 8.1 (Horton et al. 2015).

Box 8.1 (cont.)

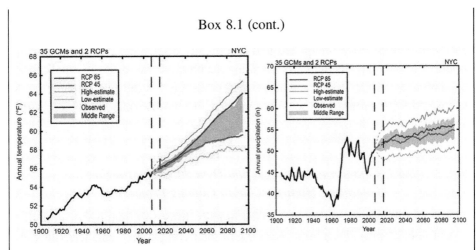

Figure 8.1 Combined observed (black line) extending to 2010 and projected (grey lines beyond 2010) temperature (right) and precipitation (left) for the New York City area. For temperature the two thick grey lines around the shaded area represent the average of the 35 GCMs for each of the RCPs (4.5 and 8.5). For precipitation, the two thick grey lines in the shaded area represent the average of the two RCPs, etc.
Source: From Horton et al. (2015)

8.6 Use of the Different Downscaling Methods

A detailed discussion of recent research comparing the different downscaling methods is provided in the "added-value" chapter. Here we review some existing documents on the use of the different downscaling methods, and, Table 8.1 provides an overview of the characteristics of the two different fundamental approaches. A discussion of the various types of ESDM and RCMs and the variety of archived datasets available from these types of models is discussed in Chapters 4 and 5.

The IPCC Task Group on Data and Scenario Support for Impacts and Climate Analysis (TGICA) has provided some recommendations in various guidance documents over time, including one for use of RCM results (Mearns et al. 2003) and for statistical downscaling (Wilby et al. 2004). Ekström et al. (2015) provide more recent guidelines and emphasize the concepts of climate realism (model skill) and physical plausibility of change in considering the use of different methods. They pose a series of questions about the nature of the planned use of the climate information that overlaps with many of our discussions in this document. However, they avoid providing overly detailed recommendations, as do we. Vano et al. (2018) provide some guidance for water resources managers in terms of "do's" and "don'ts" for using future climate information, such as "do identify all relevant uncertainties." Maraun and Widmann (2018) briefly discuss the use of downscaling in practice (chap. 18) and specifically discuss how to select suitable (global) climate models and downscaling techniques for a given application. Kotamarthi et al. (2016) in a white paper produced for the US Department of Defense SERDP program, discusses many of the more expanded topics in this current volume.

8.6.1 Comparing Statistical and Dynamical Approaches to Generating High-Resolution Climate Projections

ESDMs and RCMs and other dynamical models differ in a number of important ways. Table 8.1 provides a brief overview of their comparative characteristics. An important one that can influence the choice of method is the number of variables output by the methods. In general, most ESD methods produce only temperature and precipitation (although other variables, such as wind and humidity have been occasionally downscaled). In the case of RCMs and other dynamical methods, many different variables are produced at multiple vertical levels through their atmosphere (Chapter 4, Table 4.1 for a list). While not all of these variables are always made available for use to users, the most important ones (temperature, precipitation, wind, humidity, solar radiation) are generally made available on a daily basis.

Another important and practical difference between the methods is *efficiency*, which largely results from the different levels of computational resources needed to create simulations. Some ESDMs can bias-correct and downscale hundreds of years of GCM simulations in the same time it would take to run a few decades of RCM simulations or other dynamical downscaling approach. For this reason, most of the datasets of high-resolution climate projections that have been generated by ESDMs use methods and models that are flexible and rapid, capable of developing climate projections based on multiple scenarios and daily GCM simulations that correspond to the spatial and temporal resolution of a gridded or station-based dataset for a city, a region, or a country.

Another contrast between ESDMs and dynamical models concerns *flexibility*. ESDMs can generate output at the spatial scale of nearly any observational dataset, from a gridded dataset to an individual point source, as long as relevant and reliable observations of that variable at that scale have been collected over a sufficiently long time to be climatologically relevant. To generate similar downscaled products with RCMs, additional simulations will have to be performed for each spatial resolution and this increases the computational burden of producing the required output.

ESDMs differ from RSMs in that their output is by definition *bias corrected* such that the statistics of weather over a climatological period produced by an ESDM match those of the observations used to build the ESDM over the same period. In other words, for the historical training period, the statistical properties of ESDM output, such as seasonal temperature, or the frequency of days per year with more than a certain amount of precipitation in twenty-four hours, will be nearly identical – to within the goodness-of-fit of the statistical model – to that of the observations over the same historical period. However, given the needs of user communities, it has now become customary to provide bias-corrected results for dynamical downscaling model output (Mearns et al. 2015).

ESDMs are *limited by observations* in at least three different ways. First, it is only possible to develop projections for variables that have already been observed for a number of years, and at the scale at which they were observed. Second, statistical methods assume the observational data are a perfectly accurate representation of actual conditions. In reality, numerous factors – from observer error to long-term creep in measurement equipment – can bias observations relative to reality. Third, ESDMs can be sensitive to errors in the observational data, particularly at the extremes. This can require extensive quality control and error checking before the observations are used. Dynamical downscaling requires observations in order to validate the model results for the current period, but not to generate the simulations. Yet observations are still important for tuning dynamical models and testing model components.

Box 8.2
Case Study: UKCP2018. Climate Scenarios for the United Kingdom.

The UKCP18 scenarios are fundamentally based on a single coupled atmosphere-ocean model, HadGEM3-GC3.05, which has a high-resolution spatial grid of about 60 km in mid-latitudes (Murphy et al. 2018). A three-strand strategy was used for UKCP18 to generate information: The three different strands are provided to meet different needs of the wide-ranging user community. Strand 1 updates the probabilistic scenarios developed for UKCP09 for UK and surrounding regions. The probabilistic projections are designed to provide a measure of the uncertainty from emission scenarios and scientific uncertainty in the models. Strand 2 consists of fifteen parameter permutation experiments (PPE) simulations (1900–2100) plus thirteen simulations from other CMIP5 models. These simulations are from the GCM and are at 60 km resolution and are designed to provide the user with projections of changes in climate for the world. Strand 3 consists of a twelve-member downscaled simulations at 12 km

Continued

Box 8.2 (cont.)

horizontal resolution. These projections are produced by using a high-resolution regional scale climate model from the HadGEM3 over Europe (1980–2080). Strand 3 is useful for applications requiring analysis at local to regional scales. A second phase of Strand 3 is planned whereby the 12 km RCM simulations will drive additional RCM simulations at 2.2 km horizontal resolution using a convection resolving model. It has been demonstrated that resolving convection can have an important effect on determination of, for example, future precipitation extremes (Kendon et al. 2014). Regarding concentration scenarios, Strand 1 uses RCP 2.6, 4.5, 6.0, and 8.5. Strands 2 and 3 focus primarily on the higher RCP 8.5 scenario – see Figure 8.2.

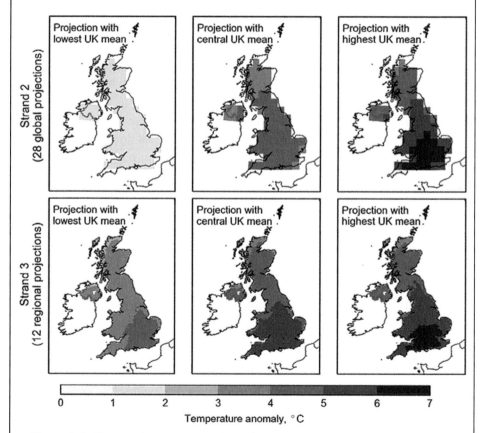

Figure 8.2 Changes in summer surface air temperature for 2061-80 relative to 1981–2000 for RCP 8.5, from individual projections selected from Strands 2 and 3. Source: Murphy et al. (2018)

Perhaps the most critical difference between dynamical and statistical models is the issue of *stationarity*. In this context, the term refers to the ability of the model to simulate how different physical processes that determine climate at a given location, and their relationship to each other, may change over time. The question of stationarity is particularly relevant to the relationship between the larger-scale processes that are resolved by a global model that may change over time, vs. the finer-scale processes that are hard-wired into an ESDM.

ESDMs assume that the relationship between the large-scale predictor and the unresolved local climate processes remains the same over decadal time scales. However, if climate change dynamically alters these physical processes relative to their present-day observed behavior, a statistical method may not be able to simulate these changes, and future projections will be biased in the direction of past rather than future regional dynamics. Studies have shown this assumption may not always be justified, particularly for precipitation (Fowler and Blenkinsop 2007). This underscores the importance of evaluating ESDMs for their ability to reproduce both observed climatology as well as long-term trends.

Finally, there have been many detailed comparisons between statistical and dynamical downscaling (e.g., Giorgi and Mearns 1991; Christensen et al. 2007) which discuss the characteristics presented in Table 8.1.

Table 8.1 *Fundamental comparisons of dynamical and statistical downscaling techniques*

Characteristic	Dynamic downscaling	Empirical-statistical downscaling
Efficiency	Need for computational resources and expertise in running the models makes it inefficient for producing downscaled products from many GCMS and scenarios	Quickly create downscaled output for a large number of GCMs and scenarios
Versatility	Generates many climate variables in one simulation	Often limited to temperature and precipitation
Flexibility	Different spatial resolutions need separate simulations	Highly flexible and can be used to quickly generate downscaling at a new resolution
Bias Correction	May require additional bias correction	Method incorporates bias correction
Observational Data Limitation	Does not depend on the availability of observational data at the downscaled location	Limited to locations with observational datasets
Stationarity	Not necessary to assume stationarity of scale resolved processes	Assumes that the statistical relationship between global climate model and observations holds in the future

Another important difference between ESDMs and RSMs is that of the computational requirements. Producing dynamical downscaling simulations requires a high level of atmospheric modeling skill and computational resources. Here, we generally assume that the user will choose from previously generated climate projections, rather than developing new downscaled climate projections. It is only recently, through the CORDEX program (Giorgi and Gutowski 2015) that there have been substantial numbers of simulations with RCMs at resolutions of at least 50 kms (or higher) throughout the world. Table 8.2 summarizes the regional model simulations available through CORDEX for the various continents. The largest number of simulations are available for Europe, followed by Africa. Details on the simulations may be found at the CORDEX website (www.cordex.org/). Currently the available simulations are driven by CMIP5 GCMs, but there are plans to produce simulations with the upcoming CMIP6 GCMs. Some of the simulations have been bias corrected (see the web site for more details).

The application of downscaled climate products should consider uncertainty and interpretation:

- Conducting subregional climate change impact assessments and adaptation planning:
  o Consider whether the decisions/situations at hand really require downscaling. For some assessments, monthly or annual scenarios derived directly from GCM outputs may be sufficient; in some cases, climate information will be used in conjunction with rather broad-scale indicators, and no value would be added by downscaling climate projections.
  o Investigate whether projections at appropriate scales already exist. Numerous regional climate change assessment projects provide downscaled data and/or tools that are freely available; however, their relevance for the problem at hand and their limitations would still need to be assessed.
  o If no projection exists and the project requires downscaled information, then a new downscaled dataset is necessary. In this case, explicitly include and investigate appropriate methods to downscale and contextualize the climate change information for the area of interest given time and resource constraints.
- Assessing uncertainty.
  o As discussed, use outputs from a variety of models (GCMs at global scale or RCMs at regional scales) and different RCPs, if possible, to quantify the degree of uncertainty in the projections. It is common, for example, to use at least two different RCPs, such as a high and a low scenario (e.g., RCP 8.5 and 4.5).
  o Ideally, using downscaled outputs from separate ESDMs, separate RSMs or a combination of ESDM and RSM should be considered where feasible

Table 8.2 *Currently available CORDEX simulations*

Domain	No. of RCM	No. of GCM	ERA interim	RCP 2.6	RCP4.5	RCP8.5	Total no of simulations	Available on ESGF	Available through POC	Available in other ways
Africa	12	15	13	12	36	37	98	64	98	?
Antarctica	8	10	8	2	11	13	35	?	?	?
Arctic	11	5	23	-	10	26	59	34	29	6
Australasia	3	11	5	-	20	24	49	-	11	38
Central Asia	1	3	1	-	3	3	10	-	10	
East Asia	2	5	2	-	5	6	13	-	13	
SEACLID[a]	8	12	3	-	15	15	48	?	?	?
South Asia	6	12	4	6	27	24	61	45	16	
EURO-CORDEX	11	11	28	20	31	34	113	61	52	
EUR-44[b] EUR-11[c]			23	18	24	40	105	49	56	
MED-CORDEX[d]	14	6	44	3	11	20	78	-	18	60
MENA[e]	6	7	8	2	12	12	34	15	19	
North America	10	5	16	1	7	24	52	9	36	6
Central America	4	10	7	6	3	12	28	23	4	1
South America	7	8	7	12	18	18	55	13	?	?

[a] SEACLID is Southeast Asia Downscale;
[b] spatial resolution of 50 km;
[c] spatial resolution of 12.5km;
[d] Mediterranean downscale;
[e] Middle East and North Africa downscale

(for example when such downscaled products are available from existing efforts). While doing this would be ideal, in actuality, using several different downscaling methods is rarely implemented.

- Using/interpreting published results:
  - o Be wary of very high-resolution maps. To produce such maps, downscaling is required, and it is important to research the process to determine how the downscaling was accomplished and determine limitations and uncertainties.
  - o Keep in mind that uncertainties arising from different sources are inherent to the projection process. Therefore, results should not be taken at face value,
  - o especially when no explicit uncertainty assessment is provided; alternatively, they should be construed as broad indicators of potential changes and impacts, dependable at larger spatial and temporal scales even if presented at fine scales.

8.7 Recommendations Based on Particular Variables, Questions Asked, and Physical Characteristics of the Region

Tables 8.3 and 8.4 provide general recommendations based on certain characteristics of the potential use of the downscaled information. Color coding is used in both tables to indicate the degree of appropriateness of each method, i.e., warm colors (red, yellow) are used to represent caution, and green is used to indicate suitability of the model. In Table 8.3 typical downscaling methods are listed along with the variables available from the models, the frequency of the output, and spatial resolutions. In the second part of that table, questions are asked (and answered) regarding the adequacy of the method for using different time averages of variables, seasonal mean values, extremes, and for analyzing phenomena (e.g., storms, hurricanes). Table 8.4 focuses on the recommended use of downscaled model products for regions with specific geographic features. This table is not designed to specify which would be a "better" or "best" product to use, but rather to suggest the general suitability of using a particular downscaled product as a function of meteorological scales of motion (Figure 2.6) that are relevant for the regional context

8.7.1 Useful vs. Usable

There remain important gaps between what scientists produce as useful information and what users recognize as usable information for decision making (Lemos et al. 2012). Lemos et al. (2012) argue that usability concerns three factors: users' perception of information fit; the level and quality of interaction between producers and users; and how the new (climate) knowledge interacts with other kinds of knowledge currently used by the users. Obviously in this volume, we are not providing the needed interactions between users and producers, but this is why we encourage users to work with individuals familiar with climate modeling, downscaling products, and some context for the actual use of climate information. For example, it might be valuable to convene a small advisory group or team of experts to jointly consider and advise on a preferred downscaling approach and ideally the collective review of any work product. This peer approach can help ensure robustness in the approach and confidence in the advancement of findings.

Another issue regarding usability is the data format used for providing the data. Much climate model output is provided in NetCDF format, which is a format not in general use by most user communities. However, it is now relatively common for "translation tools" to be provided on major websites to allow downloading data in more user-friendly formats (e.g., see the EURO-CORDEX web site).

Moreover, there is more information available discussing good practices for climate model output use (e.g., Kreienkamp et al. 2012; Roessler et al. 2017). It is important to give an early consideration to the potential need for and sources of external expertise to aid in the downscaling process, comparison of data, as well as the interpretation and use of results, and to plan accordingly (e.g., project scope, timeline, and budget).

8.7.2 Climate Services and Web Portals

There is an abundance of web sites that provide information about future climate change both on a global and regional basis. But there is a lack of uniformity on recommendations for use, specific guidance for use, how the data are presented, and whether there are any guarantees regarding the quality of the data (Hewitson et al. 2017). While there are various Climate Services organizations that address the issue of providing information on future climate (Brasseur and Gallardo 2016), there is lack of uniformity on what is actually provided. Hewitson et al. (2017) examined and assessed forty-two climate information websites, and found a wide variety of data provided. Most fundamentally rely on the global model suites of simulations associated with the IPCC (in particular CMIP3 and 5). Most provide global coverage, but an increasing number (partially as a result of CORDEX) focus on specific regions and include various downscaling methods. The diversity of information provided is largely a function of the fact that the development of climate projections for decision making really remains an evolving research area even though the need for these scenarios have been discussed for a long time (Mearns et al. 2001). Swart et al. (2017) researched the question of what are the characteristics of successful climate information portals and what are potential pitfalls, focusing particularly on Europe (e.g., CLIPC,

Climate Information Portal for Copernicus). They considered both portal design and the process of engaging users. They found that different user groups have different perceptions of strengths and weaknesses of portals such as CLIPC. But all user groups emphasized the importance of clear and comprehensive metadata and indicated that trust in the portals depends on the scientific quality of the data. Recommendations for future portal development included taking account more explicitly of the wide diversity of users and their needs, systematically managing

Table 8.3 *Evaluation of available downscaling models and output and their limitations*

Descriptions	GCMs	Empirical Statistical Downscaled Datasets							RCM Datasets		
		Multiple Linear Regression	Empirical Quantile Mapping	Bias Correction	Paramteric Quantile Mapping	Weather Geneerator	Constructed Analogues	KDDM	CORDEX	SERDP	NARCCAP
Names	Multiple Models	Many	BCSD,EQDM, CDFt	MBC	ARRM V1	SDSM	CR, LOCA	LDDM	Multiple Models	WRFV3.2	Multiple Models
Source	CMIP3/5		Maurer/Wood	Sachindra	Stoner Hayhoe	Vilby	Hidalgo	McGiniss	Gutowksi Mearns	Kotamarthi	Mearns
Format	NetCDF Output		NetCDF Output		NetCDF Output	PC Code			NetCDF	NetCDF	Netcdf
Temporal Res [IN]	Daily	Monthly	Monthly	Monthly	Daily	Monthly	Daily	Daily	3h/6h	6h	3hr/6hr
Temporal Res [OUT]	daily	Monthly	Daily	Monthly	Daaily	daily	Daily	Daily	hours to daily	3hourly	daily
Spatial Resolution	1 degree to 2.5 degrees	Individual stations	GRID - 1/8th degree	same as observation	GRID - 1/8th degree and Individual stations	Individual stations	Grid 1/8 degree	NARCCAP Grid	GRID (50km amd 20km)	GRID(12 km)	Grid 50km
Output Variables	Many	T(avg)	T(max)	T(max)	T(max)	T(max)	T(max)	T(max)	53	80	53
		Pr	T(min)	T(min)	T(min)	T(min)	T(min)	T(min)			
			Pr	Pr	Pr	Pr	Pr	Pr			
					RH (max/min)						
APPLICATIONS											
Can I use the absolute values that come from these sources, or do they have to be bias-corrected? Is this method/data adequate for …											
annual and seasonal mean temperature and/or precipitation?											
annual temperature and precipitation extremes?											
Daily mean precipitation											
decadal temperature and precipitation extremes?											
hurricanes, winter storms, and other types of large-scale extreme weather events?											

Note: Many of the empirical downscaling methods use a combination of methods for generating downscaled output. For example, weather generators are frequently used with the spatial disaggregation methods to generate time resolved temperature and precipitation. Greenish – Can be used / evaluated for this application; Yellow – Use with caution/evaluated under certain conditions; Red – use with extreme caution/untested or unavailable.

data quality, paying sufficient attention to well-developed guidance for selection and interpretation of data, and inclusion of means of transforming data to formats most used by the various user groups.

8.8 Conclusion

While we cannot recommend a single approach even for any given application, it is feasible to construct a set of options that could be desirable given the current state of knowledge. A general set of rules to follow that we can suggest is to: (a) identify

Table 8.4 *Recommendation table on the use of climate datasets based on regional features*[a]

Scale	Statistical Downscaling Methods						Dynamic Downscaling		GCM	
	Delta Correction	Equidistant Quantile Mapping	Bias correction	Parametric Quantile Mapping	Constructed Analogues	Wx generator	CORDEX (50, 20, 12 km)	SERDP (12km)	CMIP5	CMIP6
Planetary scale: ~3,000 km or more, weeks to months (general circulation structure, jet stream position)										
Synoptic scale: 100–3,000 km, days to weeks (highs and lows, midlatitude cyclones, monsoons, atmospheric teleconnections)										
Mesoscale: 10– hours to days (katabatic winds, weather fronts, mesoscale convective systems, tropical cyclones, sea breeze circulations)										
Local Scale: 1–10 km, hours to minutes (supercell thunderstorms, tornadoes, gust fronts, air mass thunderstorms, mountain-valley winds, mountain snowfall)										
Coastal Region (< 100 km from the coast)										
Complex Terrain (Mountains)										

[a] *The scales of atmospheric motion correspond to Figure 2.4 in Chapter 2. Green – can be used / evaluated for this application; Yellow – use with caution/evaluated under certain conditions; Red- use with extreme caution/untested or unavailable*

a set of criteria that can be used to broadly classify the problems a user would encounter; (b) for each of these criteria, address the suitability of using a particular method or an appropriate set of methods. Tables 8.3 and 8.4 are designed to provide general guidance for choosing a method based on the problem spatial scale and data needs. The color coding in Tables 8.3 and 8.4 should be considered a qualitative ranking and an expert judgment based on our collective knowledge.

9

The Future of Regional Downscaling

What lies in the future of regional downscaling, and how will future developments advance the information available for impacts and policy-making analyses? Models and methods are constantly under development; what might users expect to become available over the next decades as computing resources reach and extend beyond exascale (over 100 times faster than the current fastest computers), and GCM resolutions reach regional scales? What uncertainties will remain to be resolved, even with these advances? By its nature, much of this discussion is somewhat philosophical, and this chapter does not include real-world case studies. Instead, it highlights specific ways assessments may be affected by future advances and discusses key areas of future development.

9.1 A Look at the Future

Over the next several decades, climate change and its myriad of consequences will continue to unfold and likely accelerate, increasing the demand for climate information for impact assessments and policy considerations. Society will need to respond and adapt to impacts far beyond those currently observed today, from rising sea level to large-scale changes in ecosystems and agriculture and the potential for severe water scarcity in many regions. Climate change will also continue to affect the likelihood and severity of extreme weather and climate events, increasing the damages associated with these events unless rapid and comprehensive steps are taken to reduce the vulnerability and the exposure of people, services, and infrastructure. We already recognize that historical records are no longer a reliable predictor of future conditions and that we have entered a new period of rapid change that far exceeds, in both magnitude and speed, anything human civilizations have experienced in the past. Driven by these growing needs, we anticipate that the scientific tools for studying these influences will evolve in several important ways.

The first improvement on the horizon is computing power. As it increases, it allows for much higher resolution GCMs and RCMs, which in turn are able to explicitly resolve, rather than parameterize, more physical processes. The fastest supercomputers currently available operate at the petascale, roughly a quadrillion (10^{15}) calculations per second. In the last few years, petascale computing has become the primary platform for the complex models of the Earth's climate system. Over the next few years, exascale supercomputers, 1,000 times faster than petascale, are expected to become available. These computers will significantly expand possible spatial and temporal resolution of the models as well as allowing for a larger number of simulations and a better characterization of both structural and parametric uncertainty, including natural variability.

Better characterization of and, in some cases, reduction of the uncertainties discussed in Chapter 7 and elsewhere is the second main improvement expected in the future. First, time will narrow the likelihood of various scenarios, particularly if emissions exceed the cumulative carbon limits associated with lower 1.5 °C or even 2 °C targets and/or if clean energy development accelerates and fossil fuels begin to be phased out at a more rapid rate than predicted by the higher RCP7.0/SSP370 or RCP8.5/SSP585 scenarios. Second, greater computing power will allow for more simulations, which in turn will provide better understanding of the role of natural variability in moderating future change, particularly over shorter time scales and smaller spatial scales. At the same time, new data are constantly being collected and new experiments conducted, from laboratory studies to satellite, and other observational datasets that improve understanding and treatment of physical, chemical, and biological processes in the global and regional models. This in turn has the potential to reduce both structural and parametric uncertainty.

In this chapter, we expand on these expected developments and discuss how they will affect global climate models, RCMs and ESDMs, how they will alter the need for data and information from downscaling, and how user interfaces may evolve as well.

9.2 Future Directions for Global Modeling

The US National Academy of Sciences panel report, *A National Strategy for Advancing Climate Modeling* (NAS 2012), provided detailed recommendations on how the science community can advance the development of climate models while also to making the resulting analyses and projections more useful to those who use the output from these models. Their recommendations include to develop training, accreditation, and continuing education for "climate interpreters" who will act as a two-way interface between modeling advances and diverse user needs.

GCMs are some of the most computationally intensive pieces of code on the planet. As computational abilities continue to improve, so too will the resolution of the global and regional modeling tools being used to study climate change. Most GCMs are optimized for central processing units (CPUs). Rapid increases in processing capacity for the computational architecture of the next generation of high-performance computing, however, will likely be realized through a combination of CPUs and graphics processing units (GPUs). Thus, adapting GCM and RCM codes to fully leverage the GPUs and accelerated CPUs, particularly if much of the acceleration comes from GPUs, will require a substantial effort over the next few years.

The use of artificial intelligence and machine learning as a tool for conducting geophysical science at all spatial scales is also advancing (Bergen et al. 2019). A new generation of scientific computing facilities is being designed for data-intensive computing and integrating data with simulations using novel approaches. Climate models are under development to explore the data-simulation integration methodologies in combination with machine-learning approaches (Schneider et al. 2017). Development of deep neural networking (DNN) algorithmic-based methods for climate-process-model emulation are being tested (Krasnopolsky et al. 2013; Wang et al. 2019a). Emulators for entire climate models, built using DNN are also proposed (Scher 2018). These emerging applications of ML/AI have the promise of speeding up the development cycle of climate models and bring new data and process understanding to bear on the model performance quickly. It is likely that these advances, and in particular the development of whole model emulators, could eventually lead to the development of purpose-built climate models based on existing GCMs and suited for a particular stakeholder need.

A third way that GCMs are continuously improving is through the addition of more and more of the processes known to affect the Earth's climate. Key additions to GCMs in recent years have included dynamic carbon cycle and expansion of land surface modules. Future additions are likely to focus on a broad range of issues, including:

- the effects of urban environs on energy flow and albedo
- improved surface hydrology and coastal wave models
- better and more highly resolved treatment of atmospheric convection and its effects on clouds
- the addition of important structural aspects of the climate system, including carbon and methane emissions from thawing permafrost and the processes that are accelerating ice melt in Greenland and Antarctica
- improved integration of sea ice, including, for example, the ability to model how it thins and fragments as a result of the warming air and oceans

- enhanced treatment of weather processes to make GCMs more akin to the models used for shorter-term weather prediction at hourly to daily time scales, while continuing to account for all of the processes affecting climate over longer time scales

A fourth way that GCMs are constantly improving is in their spatial and often temporal resolution as well. However, all of the additions to the models listed will make them more computationally intensive, thus slowing them down and reducing the number of studies that can be done with these complex modeling tools. Thus, there is a constant tension within the science community as to whether to put more of the computation into higher model resolution, which itself can improve the model capabilities (though higher resolution can also expose errors in the current treatments of sub-grid scale processes) or to put more effort into enhanced treatment of surface or other processes that can also affect the capabilities and accuracy of the model. Generally, some combination of both occurs, with large modeling groups also doing extensive testing after each change is made to the model, to ensure they understand the effect of altering, for example, the convective parameterization of clouds or adding urban areas to the land surface module. For example, the US National Center for Atmospheric Research, home to several state-of-the-art GCMs, has been doing extensive tests with their CESM model at 0.25 degree (25 km) resolution globally with a 0.1 degree (10 km) ocean while also examining the effects of alternative approaches to grid structure. As discussed in earlier chapters, the CMIP model intercomparisons provide a standardized way to analyze and compare the latest versions of the global (and through CORDEX, now regional) climate models. CMIP6 is now well underway; where CMIP7 and beyond will take us in the further testing of the next generation of models is still unclear at this time.

Finally, a fifth aspect in which climate modeling needs to be improved is by building climate models that deliver much more predictive information for users. As Smith et al. (2014) discuss, a key step is to enable models to be built that include much more robust estimates of uncertainty, which in turn guides where scientific and computational resources need to be directed in order to reduce uncertainties further. Combining adaptive hierarchical modeling frameworks with assessments of the uncertainty in model formulations and projections will enable more targeted explorations of specific issues and allow urgent questions to be answered in a much more timely and reliable way.

Process-based knowledge at the local level, e.g., the effects of cities and their interactions with climate, are likely to greatly influence the further development of GCMs. Cities and their associated urban areas have a much larger impact on the environment at both local and global scales than their spatial footprint on the Earth's surface. At this time, more than half of the world's population lives in urban areas (80 percent in the United States), and this proportion is projected to climb to 70 percent by 2050. Therefore, it is essential that numerical models used to study the physics, chemistry, and biology affecting the Earth system represent the effects of urban areas on climate and the effects of changing climate on urban areas.

9.3 Future Directions for Regional Modeling

Just as GCMs have transitioned into Earth System Models over the last five years by incorporating dynamic carbon cycles, a similar transition is occurring with RCMs. The development of interactively coupled Regional Earth System Models (RESMs) is currently underway (Figure 9.1) and a few RESMs now exist for some regions. Like global models, however, they tend to be incomplete in their representation of all potentially relevant processes (Giorgi and Gutowski 2015;

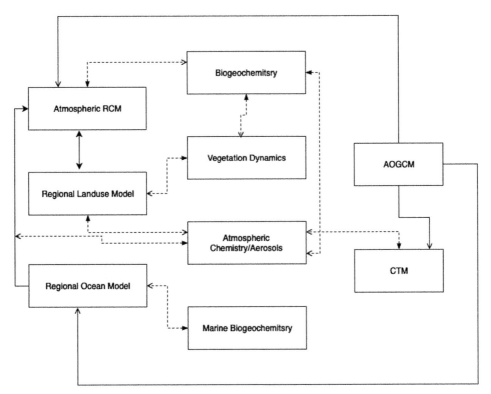

Figure 9.1 Depiction of a coupled RESM framework and the interactions across its different components and global climate model drivers (CTM: chemical transport model). Arrows indicate the flow of information. Solid line: interaction with driving global models; broken lines: interaction inside the RESM.
Source: Based on Giorgi and Gao (2018)

Giorgi and Gao 2018), though some RESMs now include multiple Earth system components, such as atmosphere, oceans, sea ice, cryosphere, hydrology, and land and/or marine biogeochemistry.

Much more work and analysis, including intercomparison studies, are needed to assess RESMs in different regional settings and to evaluate the importance of their representation of coupled processes. In particular, the inclusion of an interactive biosphere has been limited so far, but the interest in this aspect of coupled modeling is indeed growing in view of the role it will play especially within the context of future climate change, which may lead to pronounced changes in natural ecosystems (Giorgi and Gao 2018).

As with the discussion about global models, the next major challenge in RESM modeling is better consideration of human factors. Human activities, such as land-use change and greenhouse gas and aerosol emissions, are currently considered in most model experiments as external players in the climate system, either as forcings or as receptors (e.g., impacts). However, there is a two-way interaction between human societies and the natural environment, whereby on the one hand human populations may adapt in response to climatic, environmental, and socioeconomic stresses, and on the other hand adaptation policies to respond to climate change may in turn affect climate (Giorgi and Gao 2018). In an era where humans are now a key component of the climate system, these processes need to be included in ESMs. RESMs can be an optimal test-bed for this model development because these models can focus on specific regional interactions, and the lessons learned from this exercise can eventually be extended and generalized to the global ESMs.

9.4 Future Directions for Empirical-Statistical Downscaling

Recent advances in the field of ESDMs focus on four areas. The first is the inclusion of advanced statistical, signal-processing, pattern recognition, probability and machine-learning algorithms that were originally developed in other fields of research, from statistics and mathematics to computer science and engineering. These techniques offer significant potential to expand the field of ESDM and the ability of ESDMs to characterize large, three-dimensional, multi-variate datasets beyond the rather rudimentary mono-variate regression or distribution-based approaches that have characterized them in the past. Through allowing the ESDM to sample a greater range of information from the GCM and potentially the observations as well, such techniques also have the potential to increase the added value and the stationarity of the model.

The second notable area of advance is the increasing availability of long-term, high-resolution observations that can be used to train an ESDM. From satellite data to affordable button sensors to mobile observing networks and high-density mesonets, the amount of data available for downscaling is increasing yearly. Many of these datasets have reached or are near reaching the length (at least twenty to thirty years of continuous observations) needed to provide a sufficient sample for downscaling. Additional data collection in remote areas is further enhancing ESDM ability to provide high-resolution climate projections for regions and locations where it was not previously possible.

The third area is the developing focus on innovative and standardized evaluations that compare multiple ESDMs, as described in Chapter 5. Evaluations such as the ENSEMBLE project in Europe or the "perfect model" approach in North America provide key insights into the limitations and abilities of specific ESDMs for specific regions and variables. This in turn provides essential and valuable input to users in the form of quantitative guidance on the biases associated with ESDMs for specific locations.

And finally, the fourth area is the emerging focus on process-based evaluation of ESDMs, a powerful tool that has the potential to significantly increase the scientific robustness of these models. One example, the fact that current ESDMs are not yet able to account for the influence of melting snow on temperatures in high elevation locations, was already highlighted in Chapter 5. As another example, Lazante et al. (2018) identified two potential mechanisms that cause snow in the Great Lakes: large-scale cyclonic systems and local-scale lake effects. Suppose that an ESDM or an RCM performed well in simulating precipitation from large-scale systems, but not as well with that induced by lake effects. In the historical period, statistical downscaling might be based on a compromise relationship, but yield acceptable results. In the future, as a result of decreasing baroclinicity, the importance of large-scale precipitation might decline. But with an increase in lake temperatures and a longer ice-free season, the relative importance of lake-effect snows might increase. The combination of these two changes might result in much poorer statistical downscaling results in the future, something not readily apparent from historical diagnostics alone. The advanced evaluation frameworks that are being developed will allow ESDMs to be tested for these types of physical, process-based limitations and, if identified, it is conceivable that ESDMs could be modified or a post-processing step be added specifically to address the geographic challenges of specific locations such as these.

Finally, additional development is needed in statistical downscaling to more effectively support climate change applications. Some of the issues to consider are the development of ESDM techniques to downscale severe weather events; the translation of new insights about uncertainty into practical guidance about risk for decision making; and downscaling within probabilistic frameworks to better represent natural variability.

9.5 Will Downscaling Become Obsolete?

As computational abilities advance, will regional climate models still be needed? Global models are now regularly run at horizontal resolutions of 1-degree latitude and longitude, roughly 100 km (about 62 miles) over mid-latitudes. This is clearly insufficient for many, if not most, studies considering the impacts of climate change and resulting resilience or adaptation analyses and policy making. Some experimental global models are being run at 0.25-degree (25 km or 15.5 miles) resolutions, but these are still extremely computationally expensive, so there are only a few runs and few ensembles are available.

With advances in exascale computing, GCM resolutions of 25 km or finer will likely become commonplace. At the same time, however, regional modeling studies are heading toward runs at resolutions of 3 to 10 km or less and, for the limited areas of urban areas, to resolutions as fine as 100 meters. Computing capabilities will need to extend far beyond exascale to achieve these very high resolutions in global models for multi-decadal analyses. Thus, it is conceivable that regional models may one day not be needed; but that day still remains far in the future.

What about statistical downscaling methods – will they still be needed? As mentioned previously, ESDMs serve a key role in correcting the bias in both global and regional models relative to observations. Until models are able to be run at such fine resolution and with such greatly improved structural and parametric uncertainty that their biases at the local to regional scale are virtually negligible, ESDMs will still have an important role to play. In addition, many impact analyses rely on observations from local weather stations, and ESDMs offer a way to downscale climate projections directly to those individual station locations. In general, however, downscaling is likely headed toward a more integrated approach, where ESDMs are incorporated into RCMs such that GCMs can be both dynamically and statistically downscaled at the same time, as part of a hybrid approach that builds on the strengths of both methods to produce high-resolution, bias-corrected output.

9.6 Coupling with GIS and Other Tools

Policy makers and other users of climate projections often need tools that help translate the scientific findings into more readily digested information. Geographic Information Systems (GIS) can be a particularly powerful tool for translating scientific observations and understanding both to further the science and also to address the needs of data users. While there have been some applications to climate studies, there is much more need for GIS in climate analyses than is being done currently. Using GIS techniques and software, experts can closely monitor changes in climate and their relationships to societal impacts. Robust geospatial data and detailed visualizations support decision making in regional organizations and government agencies as they plan for the challenges ahead. By bringing together GIS and climate change studies, spatial problem solvers can seize opportunities to make a difference in the lives of future generations.

For example, cities today face unprecedented challenges from phenomena such as human migration and population growth, disruptive technological change, and increased social inequality and conflict. However, complex urban systems present unique and grand challenges that can only be solved with innovative actions and unprecedented coalitions of disciplines and stakeholders. In addition, urban growth creates cascading demands for clean energy, improved transportation systems, replacement of aging infrastructure, improved stormwater management to prevent flooding, better land-use planning, and more effective waste-management practices. Climate change is adding to these stresses.

New capabilities are needed for integrating these many capabilities to address the problems and issues faced by cities. One aspect of this is to develop an innovative framework to diagnose and tackle complex urban challenges through integrating analysis, measurement, and modeling with GIS. An innovative and transformational strategy is needed that can provide a decision support tool for cities with capabilities to evaluate which adaptive strategy can work for a city with a particular configuration. It can empower urban planners and engineers to designing more sustainable cities for the future and help with associated operations and resource-management decisions to reduce urban stress based on scientific understanding from a variety of disciplines. This can also help cities better understand and address issues like those mentioned, including the impacts resulting from climate change.

References

Abatzoglou, J. T. and T. J. Brown (2012). A comparison of statistical downscaling methods suited for wildfire applications. *International Journal of Climatology*, 32: 772–80.

Adarsh, S. and M. Janga Reddy (2018a). Developing hourly intensity duration frequency curves for urban areas in India using multivariate empirical mode decomposition and scaling theory. *Stochastic Environmental Research and Risk Assessment: Research Journal*, 32(6): 1889–902.

Adarsh, S. and M. Janga Reddy (2018b). Multiscale characterization and prediction of monsoon rainfall in India using Hilbert–Huang transform and time-dependent intrinsic correlation analysis. *Meteorology and Atmospheric Physics*, 130(6): 667–88.

Adger, W. N., S. Agrawala, and M. Mirza (2007). Assessment of adaptation practices, options, constraints and capacity. In IPCC (ed.), *Climate Change 2007: Climate Change Impacts, Adaptation, and Vulnerability*. Cambridge: Cambridge University Press: 718–43.

Adger, W. N., W. A. Nigel, and E. L. Tompkins (2005). Successful adaptation to climate change across scales. *Global Environmental Change: Human and Policy Dimensions*, 15(2): 77–86.

Alam, Shahabul and Amin Elshorbagy (2015). Quantification of the climate change-induced variations in intensity–duration–frequency curves in the Canadian prairies. *Journal of Hydrology*, 527(August): 990–1005.

Allen, E. B. and R. T. T. Forman (1976). Plant species removals and old-field community structure and stability. *Ecology*, 57(6): 1233–43.

Aloysius, N. R., J. Sheffield, J. E., Saiers, H. Li, and E. F. Wood (2016). Evaluation of historical and future simulations of precipitation and temperature in Central Africa from CMIP5 climate models. *Journal of Geophysical Research, D: Atmospheres*, 121(1): 130–52.

Anis, M. R. and M. Rode (2015). A new magnitude category disaggregation approach for temporal high-resolution rainfall intensities. *Hydrological Processes*, 29(6): 1119–28.

Annamalai, H., K. Hamilton, and K. R. Sperber (2007). The South Asian summer monsoon and its relationship with ENSO in the IPCC AR4 simulations. *Journal of Climate*, 20(6): 1071–92.

Arrhenius S. (1896). On the influence of carbonic acid in the air upon the temperature of the ground. *Philosophical Magazine and Journal of Science*, 41: 237–76.

Arrhenius, S. (1906). *Världarnas Utveckling*. Stockholm: H Geber.

Avissar, R. and D. Werth (2005). Global hydroclimatological teleconnections resulting from tropical deforestation. *Journal of Hydrometeorology*, 6(2): 134–45.

Barros, V. R., C. B. Field, D. J. Dokken, et al. (eds.) (2014). IPCC, 2014: Climate Change 2014: Impacts, Adaptation, and Vulnerability. Part B: Regional Aspects. Contribution of Working Group II to the Fifth Assessment Report of the Intergovernmental Panel on Climate Change. Cambridge and New York: Cambridge University Press.

Barsugli, Joseph J., Galina Guentchev, Radley M. Horton, et al. (2013). The practitioner's dilemma: How to assess the credibility of downscaled climate projections. *Eos, Transactions American Geophysical Union*, 94(46): 424–5.

Basha, G., P. Kishore, M. Venkat Ratnam, et al. (2017). Historical and projected surface temperature over India during the 20th and 21st century. *Scientific Reports*, 7(1): 2987.

Bedsworth, L., D. Cayan, G., Franco, L. Fisher, and S. Ziaja (2018). California Governor's Office of Planning and Research, Scripps Institution of Oceanography, California Energy Commission, California Public Utilities Commission. Statewide Summary Report. California's Fourth Climate Change Assessment. Publication number: SUM-CCCA4–2018-013.

Benestad, R. E. (2017). A mental picture of the greenhouse effect. *Theoretical and Applied Climatology*, 128: 679–88. Available at: http://dx.doi.org/10.1007/s00704–016–1732-y

Benestad, R. E., I. Hanssen-Bauer, and D. Chen (2008). *Empirical-Statistical Downscaling*. Singapore: World Scientific: 228

Benestad, R. E., D. Nychka, and L. O. Mearns (2012). Spatially and temporally consistent prediction of heavy precipitation from mean values. *Nature Climate Change,* 2 (April): 544.

Benoit, L. and G. Mariethoz (2017). Generating synthetic rainfall with geostatistical simulations. *Wiley Interdisciplinary Reviews: Water,* 4(2): e1199.

Bergen, Karianne J., Paul A. Johnson, Maarten V., de Hoop, and Gregory C. Beroza (2019). Machine learning for data-driven discovery in solid earth geoscience. *Science*, 363(6433). Available at: https://doi.org/10.1126/science.aau0323

BOM and CSIRO (2012). Annual Australian Climate Statement. Available at: www.bom .gov.au/climate/current/annual/aus/2012/

BOM and CSIRO (2014). Annual Australian Climate Statement. Available at: www.bom .gov.au/climate/current/annual/aus/2014/

BOM and CSIRO (2016). Annual Australian Climate Statement. Available at: www.bom .gov.au/climate/current/annual/aus/2016/

BOM and CSIRO (2018). Annual Australian Climate Statement. Available at: www.bom .gov.au/climate/current/annual/aus/2018/

Braconnot, P., S. P. Harrison, M. Kageyama, et al. (2012). Evaluation of climate models using palaeoclimatic data. *Nature Climate Change*, 2: 417–24. Available at: http://dx .doi.org/10.1038/nclimate1456

Brasseur, G. P. and Laura Gallardo (2016). Climate services: Lessons learned and future prospects. *Earth's Future*, 4(3): 79–89.

Brasseur, G. P., D. Jacob, and S. Schuck-Zoller (eds.) (2017). *Klimawandel in Deutschland:* Openn Access Publication, e-online, Springler Spektrum.

Breinl, Korbinian, Thea Turkington, and Markus Stowasser (2015). Simulating daily precipitation and temperature: A weather generation framework for assessing hydrometeorological hazards. *Meteorological Applications*, 22(3): 334–47.

Brown, Casey (2011). Decision-scaling for robust planning and policy under climate uncertainty. *World Resources Report Uncertainty Series*, 14.

Brown, M. E., J. M. Antle, P. Backlund, et al. (2015). *Climate Change, Global Food Security, and the U.S. Food System.* Available at: www.usda.gov/oce/climate_ change/FoodSecurity2015Assessment/FullAssessment.pdf

Bruyère, C. L., J. M., Done, G. J., Holland, and S. Fredrick (2014). Bias corrections of global models for regional climate simulations of high-impact weather. *Climate Dynamics*, 43(7): 1847–56.

Budyko, M. I. (1969). The effect of solar radiation variations on the climate of the Earth. *Tellus*, 21: 611–19,

Bukovsky, Melissa S., Carlos M. Carrillo, David J. Gochis, et al. (2015). Toward assessing NARCCAP regional climate model credibility for the North American monsoon: future climate simulations. *Journal of Climate*, 28(17): 6707–28.

Bukovsky, M. S., D. J. Gochis, and L. O. Mearns (2013). Towards assessing NARCCAP Regional Climate Model credibility for the North American monsoon: Current climate simulations. *Journal of Climate*, 26(22): 8802–26.

Bukovsky, Melissa S., Joshua A. Thompson, and Linda O. Mearns (2019). Weighting a regional climate model ensemble: Does it make a difference? Can it make a difference? *Climate Research*, 77 (1): 23–43.

Bukovsky, M. S., R. R. McCrary, A. Seth, and L. O. Mearns (2017). A mechanistically credible, poleward shift in warm-season precipitation projected for the U.S. Southern Great Plains? *Journal of Climate* 30: 8275–98.

Bush, E. and D. S. Lemmen (eds.) 2019. Canada's Changing Climate Report; Government of Canada, Ottawa, ON. 444p

Busuioc, A., R. Tomozeiu, and C. Cacciamani (2008). Statistical downscaling model based on canonical correlation analysis for winter extreme precipitation events in the Emilia-Romagna region. *International Journal of Climatology*, 28: 449–64.

Cammalleri, C., P. Barbosa, F. Micale, J. V. Vogt. (2017) Change impacts and adaptation in Europe, focusing on extremes and adaptation until the 2030s. PESETA-3 Project, Final Sector Report on Task 9: Droughts, European Commission, JRC Ispra.

Cash, D., W. C. Clark, F. Alcock, et al. (2002). Salience, credibility, legitimacy and boundaries: Linking research, assessment and decision making. Available at: https://doi.org/10.2139/ssrn.372280

Castro, C. L., H.-I. Chang, F. Dominguez, et al. (2012). Can a regional climate model improve the ability to forecast the North American Monsoon? *Journal of Climate*, 25 (23): 8212–37.

Charney, Jule, A. Arakawa, D. J. Baker, et al. (1979). Carbon dioxide and climate: A scientific assessment. A report of the ad hoc Study Group on Carbon Dioxide and Climate to the Climate Research Board of the National Research Board. Available at: www.bnl.gov/envsci/schwartz/charney_report1979.pdf

Cheung, W. W., V. W. Lam, J. L. Sarmiento, et al. (2009). Projecting global marine biodiversity impacts under climate change scenarios. *Fish and Fisheries*, 10: 235–51. DOI:10.1111/j.1467-2979.2008.00315.x

Chong-Hai, X. and X. Ying. (2012). The projection of temperature and precipitation over China under RCP scenarios using a CMIP5 Multi-Model Ensemble. *Atmospheric and Oceanic Science Letters*, 5(6): 527–33. DOI:10.1080/16742834.2012.11447042

Christensen, Jens Hesselbjerg, Bruce Hewitson, Aristita Busuioc, et al. (2007). Regional climate projections. In *Climate Change, 2007: The Physical Science Basis. Contribution of Working group I to the Fourth Assessment Report of the Intergovernmental Panel on Climate Change, University Press, Cambridge, Chapter 11*: 847–940.

Ciscar, J.-C., A. Iglesias, L. F. László Szabó, et al. (2011). Physical and economic consequences of climate change in Europe. *Proceedings of the National Academy of Sciences of the United States of America*, 108(7): 2678–83.

Crowley, T. J. (1990). Are there any satisfactory geologic analogs for a future greenhouse warming? *Journal of Climate*, 3(11): 1282–92.

CSIRO and Bureau of Meteorology (2015). Climate Change in Australia Information for Australia's Natural Resource Management Regions: Technical Report, CSIRO and Bureau of Meteorology, Australia.

Cui, M., H. V. O. N. Storch, and E. Zorita (1995). Coastal sea level and the large-scale climate state a downscaling exercise for the Japanese islands. *Tellus, A*47(1): 132–44.

Dai, A. (2012). Increasing drought under global warming in observations and models. *Nature Climate Change*, 3(August): 52.

Department of Environmental Affairs (2017). South Africa's 2nd Annual Climate Change Report: Department of Environmental Affairs, South Africa. Available at: www .environment.gov.za/sites/default/files/reports/southafrica_secondnational_climatech nage_report2017.pdf

Deser, C., A. S. Phillips, M. A. Alexander, and B. V. Smoliak (2014). Projecting North American climate over the next 50 years: Uncertainty due to internal variability. *Journal of Climate*, 27: 2271–96. Available at: http://dx.doi.org/10.1175/JCLI-D-13-00451.1

Deser, C., S. Solomon, R. Knutti, and A. S. Phillips (2012). Communication of the role of natural variability in future North American climate. *Nature Climate Change*, 2: 775–9. Available at: http://dx.doi.org/10.1038/nclimate1562

Dessai, Suraje and Mike Hulme (2004). Does climate adaptation policy need probabilities? *Climate Policy*, 4(2): 107–28.

Dettinger, M. D., D. R. Cayan, M. K. Meyer, et al. (2004). Simulated hydrologic responses to climate variations and change in the Merced, Carson, and American River Basins, Sierra Nevada, California, 1900–2099. *Climatic Change,* 62: 283–317.

Di Luca, A., R. de Elía, and R. Laprise (2012). Potential for added value in precipitation simulated by high-resolution nested regional climate models and observations. *Climate Dynamics,* 38(5): 1229–47.

Di Luca, A., R. de Elía, and R. Laprise (2015). Challenges in the quest of added value of regional climate dynamical downscaling. *Current Climate Change Reports*, 1:10–21.

Dickinson, R., R. Errico, F. Giorgi, and G. Bates (1989). A regional climate model for the Western United States. *Climatic Change*. Available at: https://doi.org/10.1007/bf00240465

Dilling, L. and M. Lemos (2011). Creating useable science: Opportunities and constraints for climate knowledge use and their implications for science policy. *Global and Environmental Change,* 21: 680–9

Dixon, K. W., J. R. Lanzante, M. J. Nath, et al. (2016). Evaluating the stationarity assumption in statistically downscaled climate projections: Is past performance an indicator of future results? *Climatic Change*, 135: 395–408. Available at: http://dx .doi.org/10.1007/s10584–016-1598-0

Dosio, A. and H. Panitz (2016). Climate change projections for CORDEX-Africa with COSMO-CLM regional climate model and differences with the driving global climate models. *Climate Dynamics,* 46(5): 1599–625.

Dutton, J. A. (1995). *Dynamics of Atmospheric Motion*. Mineola, New York: Dover Publications.

Easterling, D. R., K. E. Kunkel, J. R. Arnold, et al. (2017). Precipitation change in the United States. In D. J. Wuebbles, D. W. Fahey, K. A. Hibbard, et al. (eds.), *Climate Science Special Report: Fourth National Climate Assessment*, Volume I, Washington, DC: U.S. Global Change Research Program: 207–230. DOI: 10.7930/J0H993CC

Edenhofer O., R. Pichs-Madruga, Y. Sokona et al. (2014). Technical Summary. In O. Edenhofer, R. Pichs-Madruga, Y. Sokona, et al. (eds.), *Climate Change 2014: Mitigation of Climate Change. Contribution of Working Group III to the Fifth*

Assessment Report of the Intergovernmental Panel on Climate Change. Cambridge and New York: Cambridge University Press: 1–1419.

Edwards, P. N. (2011). History of climate modeling. *WIREs Climate Change,* 2: 128–39.

Ekström, Marie, Michael R. Grose, and Penny H. Whetton, et al. (2015). An appraisal of downscaling methods used in climate change research. *Wiley Interdisciplinary Reviews: Climate Change.* Available at: https://doi.org/10.1002/wcc.339

Erfanian, A., G. Wang, and L. Fomenko (2017). Unprecedented drought over tropical South America in 2016: Significantly under-predicted by tropical SST. *Scientific Reports,* 7(1): 5811.

Evans, J. P. and M. F. McCabe, (2013). Effect of model resolution on a regional model simulation over southeast Australia. *Climate Research,* 56: 131–45.

Eyring, V., G. Flato, J.-F. Lamarque, et al. (2019). *CMIP6 Analysis Workshop.* Barcelona, Spain. Available at: https://cmip6workshop19.sciencesconf.org/data/CMIP6_CMI P6AnalysisWorkshop_Barcelona_190325_FINAL.pdf

Fahad, M. G., S., A. K., Islam, R. A. Nazari et al. (2018). Regional changes of precipitation and temperature over Bangladesh using bias-corrected multi-model ensemble projections considering high-emission pathways. *International Journal of Climatology,* 38: 1634–48. doi:10.1002/joc.5284

Famien, Adjoua Moise, Serge Janicot, Abe Delfin Ochou, et al. (2018). A bias-corrected CMIP5 dataset for Africa using the CDF-T Method – a contribution to agricultural impact studies. Earth System Dynamics. Available at: https://doi.org/10.5194/esd-9-313-2018

Felber, Raphael, Sibylle Stoeckli, and Pierluigi Calanca (2018). Generic calibration of a simple model of diurnal temperature variations for spatial analysis of accumulated degree-days. *International Journal of Biometeorology,* 62(4): 621–30.

Feser, Frauke, Burkhardt Rockel, Hans von Storch, Jörg Winterfeldt, and Matthias Zahn (2011). Regional climate models add value to global model data: A review and selected examples. *Bulletin of the American Meteorological Society* 92, (9): 1181–92.

Field, C. B., V. R. Barros, K. J. Mach, et al. (2014). Technical summary. In C. B. Field, V. R. Barros, D. J. Dokken, et al. (eds.), *Climate Change 2014: Impacts, Adaptation, and Vulnerability. Part A: Global and Sectoral Aspects. Contribution of Working Group II to the Fifth Assessment Report of the Intergovernmental Panel on Climate Change.* Cambridge, UK and New York: Cambridge University Press: 35–94.

Förster, Kristian, Florian Hanzer, Benjamin Winter, Thomas Marke, and Ulrich Strasser (2016). An Open-Source MEteoroLOgical Observation Time Series DISaggregation Tool (MELODIST v0.1.1). *Geoscientific Model Development,* 9(7): 2315–33.

Fowler, H. J. and S. Blenkinsop (2007). Linking climate change modelling to impacts studies: Recent advances in downscaling techniques for hydrological modelling. *International Journal of Climatology.* Available at: https://rmets.onlinelibrary.wiley .com/doi/abs/10.1002/joc.1556.

Fox-Rabinovitz, M. S., E. H. Berbery, L. L. Takacs, and R. C. Govindaraju (2005). A Multiyear Ensemble Simulation of the US Climate with a Stretched-Grid GCM. *Monthly Weather Review,* 133(9): 2505–25.

Füssel, H., A. Jol, A. Marx, et al. (2017). Climate Change, Impacts and Vulnerability in Europe 2016-An Indicator-Based Report. Luxembourg: EEA Report, No 1/2017, Publications Office of the European Union.

Fyfe, J. C., G. A. Meehl, M. H. England, et al. (2016). Making sense of the early-2000s warming slowdown. *Nature Climate Change,* 6: 224–8. Available at: http://dx.doi .org/10.1038/nclimate2938

Ganguli, Poulomi and Paulin Coulibaly (2019). Assessment of future changes in intensity-duration-frequency curves for Southern Ontario using North American (NA)-CORDEX models with nonstationary methods. *Journal of Hydrology: Regional Studies*, 22: 100587.

Gates, W. L. (1985). The use of general circulation models in the analysis of the ecosystem impacts of climatic change. *Climatic Change*, 7(3): 267–84

Giorgi, F. and G. T. Bates (1989). The climatological skill of a regional model over complex terrain. *Monthly Weather Review*, 117(11): 2325–47.

Giorgi, F. and E. Coppola (2010). Does the model regional bias affect the projected regional climate change? An analysis of global model projections. *Climatic Change*, 100, 787–95. Available at: http://dx.doi.org/10.1007/s10584–010-9864-z

Giorgi, F. and X.-J. Gao (2018). Regional earth system modeling: Review and future directions. *Atmospheric and Oceanic Science Letters*, 11: 189–97. DOI: 10.1080/16742834.2018.1452520

Giorgi, F. and W. J. Gutowski (2015). Regional dynamical downscaling and the CORDEX initiative. *Annual Review of Environment and Resources*, 40: 467–90.

Giorgi, F. and Linda O. Mearns (1991). Approaches to the simulation of regional climate change: A review. *Reviews of Geophysics, SCOPE*, 29(2): 191.

Giorgi, F., C. Torma, E. Coppola, et al. (2016). Enhanced summer convective rainfall at Alpine high elevations in response to climate warming. *Nature Geoscience*, 9(8): 584.

Glahn, H. R. and D. A. Lowry (1972). The use of Model Output Statistics (MOS) in objective weather forecasting. *Journal of Applied Meteorology*, 11: 1203–11

Gudmundsson, L., J. B. Bremnes, J. Haugen (2012). Scaling RCM precipitation to the station scale using statistical transformations – a comparison of methods. *Hydrology and Earth System Sciences*, 16: 3383–90. doi:10.5194/hess-16-3383-2012

Gutiérrez, J. M., D. Maraun, M. Widmann, et al. (2019). An intercomparison of a large ensemble of statistical downscaling methods over Europe: Results from the VALUE perfect predictor cross-validation experiment. *International Journal of Climatology*, 39: 3750– 85. https://doi.org/10.1002/joc.5462

Gutmann, E., T. Pruitt, M. P. Clark, et al. (2014). An intercomparison of statistical downscaling methods used for water resource assessments in the United States, *Water Resources Journal*, 50: 7167– 86. doi:10.1002/2014WR015559

Gutmann, E. D., R. M. Rasmussen, C. Liu, et al. (2012). A comparison of statistical and dynamical downscaling of winter precipitation over complex terrain. *Journal of Climate*, 25: 262–81.

Havard, M., J. Jean, M. Saddier, et al. (2015). French National Climate Change Impact Adaptation Plan 2011–2015 (INIS-FR–19-0198). France.

Hawkins, E. and R. Sutton, (2009). The potential to narrow uncertainty in regional climate predictions. *Bulletin of the American Meteorological Society*, 90: 1095–107. Available at: http://dx.doi.org/10.1175/2009BAMS2607.1

Hawkins, E. and R. Sutton (2011). The potential to narrow uncertainty in projections of regional precipitation change. *Climate Dynamics*, 37: 407–8. Available at: http://dx.doi.org/10.1007/s00382–010-0810-6

Hay, L. and M. P. Clark (2003). Use of statistically and dynamically downscaled atmospheric model output for hydrological simulations in three mountainous basins in the western US. *Journal of Hydrology*, 282: 56–75.

Hayhoe, K., J. Edmonds, R. E. Kopp, et al. (2017). Climate models, scenarios, and projections. In D. J. Wuebbles, D. W. Fahey, K. A. Hibbard, et al. (eds.), *Climate Science Special Report: Fourth National Climate Assessment, Volume I*. U.S. Global Change Research Program: 133–60. doi:10.7930/J0WH2N54

Hayhoe, K., I., Scott-Fleming, and A. M. K. Stoner, (2020). STAR-ESDM: High-Resolution Station and Grid-Based Climate Projections, 3A.3, American Meteorological Society Annual Meeting, Boston, January 13.

Hayhoe, K., J. van Dorn, T. Croley II, and N. Schlegal (2010). Regional climate change projections for Chicago and the US Great Lakes. *Journal of Great Lakes Research*. Available at: www.sciencedirect.com/science/article/pii/S0380133010000559

Hayhoe, K., D. J. Wuebbles, D. R. Easterling, et al. (2018). Our changing climate. In D. R. Reidmiller, C. W. Avery, D. R. Easterling, et al. (eds.), *Impacts, Risks, and Adaptation in the United States: Fourth National Climate Assessment, Volume II*. Washington, DC: US Global Change Research Program, Washington, DC: 72–144. doi: 10.7930/NCA4.2018.CH2

Hewitson, B. C. and R. G. Crane. (1996). Climate downscaling: Techniques and application. *Climate Research*, 7: 85–95.

Hewitson, Bruce, Katinka Waagsaether, Jan Wohland, Kate Kloppers, and Teizeen Kara (2017). Climate information websites: an evolving landscape. *Wiley Interdisciplinary Reviews: Climate Change*, 8(5): e470.

Hidalgo, H. G., M. D. Dettinger, and D. R. Cayan (2008). *Downscaling with Constructed Analogues: Daily Precipitation and Temperature Fields Over the United States*. California Energy Commission, PIER Energy-Related Environmental Research. CEC-500-2007-123.

Hijioka, Y., E. Lin, J. J. Pereira, et al. (2014). Asia. In V. R. Barros, C. B. Field, D. J. Dokken, et al. (eds.), *Climate Change 2014: Impacts, Adaptation, and Vulnerability. Part B: Regional Aspects. Contribution of Working Group II to the Fifth Assessment Report of the Intergovernmental Panel on Climate Change*. Cambridge, UK and New York: Cambridge University Press: 1327–70

Hijmans, R. J., Susan Cameron, Juan Parra, et al. (2005). WorldClim, Version 1.3. Berkeley CA.: University of California.

Hong, C., Q. Zhang, Y. Zhang, et al. (2017). Multi-year downscaling application of two-way coupled WRF v3. 4 and CMAQ v5. 0.2 over East Asia for regional climate and air quality modeling: Model evaluation and aerosol direct effects. *Geoscientific Model Development,* 10 (6): 2447–70.

Horton, R., D. Bader, Y. Kushnir, et al. (2015). New York City Panel on Climate Change 2015 Report Chapter 1: Climate observations and projections. *Annals of the New York Academy of Sciences'* 1336: 18–35. doi:10.1111/nyas.12586

Horton, Radley M., Ethan D. Coffel, Jonathan M. Winter, and Daniel A. Bader (2015). Projected changes in extreme temperature events based on the NARCCAP Model Suite. *Geophysical Research Letters,* 42(18): 7722–31.

Huang, J., Ji, M., Xie, Y., Wang, S., He, Y., and Ran, J. (2016). Global semi-arid climate change over last 60 years. *Climate Dynamics*, 46(3–4): 1131–50.

Huth, R. (2002). Statistical downscaling of daily temperature in central Europe. *Journal of Climate,* 15(13): 1731–42.

Im, E.-S., J. S. Pal, and E. A. B. Eltahir (2017). Deadly heat waves projected in the densely populated agricultural regions of South Asia. *Science Advances,* 3(8): e1603322.

Imbach, P., S. C. Chou, A. Lyra, et al. (2018). Future Climate Change Scenarios in Central America at High Spatial Resolution. *PloS One,* 13(4): e0193570.

IPCC (1996). *Climate Change 1995: Impacts, Adaptation and Mitigation of Climate Change: Scientific-technical Analysis*. Cambridge, UK and New York: Cambridge University Press.

IPCC (2007). *Climate Change 2007: Synthesis Report. Contribution of Working Groups I, II and III to the Fourth Assessment Report of the Intergovernmental Panel on Climate*

Change. Core Writing Team, R. K. Pachauri and A. Reisinger (eds.). Geneva, Switzerland: IPCC:104 pp.

INCCA (2010). *Climate Change and India: A 4x4 Assessment – A Sectoral and Regional Analysis for 2030s*. New Delhi, India: Indian Network for Climate Change Assessment, Ministry of Environment and Forests, Government of India.

IPCC (2013). *Climate Change 2013: The Physical Science Basis. Contribution of Working Group I to the Fifth Assessment Report of the Intergovernmental Panel on Climate Change*. T. F. Stocker, D. Qin, and G.-K. Plattner, (eds.), Cambridge, UK and New York, NY: Cambridge University Press: 1535 pp.

Irving, Damien B., Penny Whetton, and Aurel F. Moise (2012). Climate projections for Australia: A first glance at CMIP5. *Australian Meteorological and Oceanographic Journal*, 62(4): 211–25.

Jacob, D., J. Petersen, B. Eggert, et al. (2014). EURO-CORDEX: New high-resolution climate change projections for European impact research. *Regional Environmental Change*, 14(2): 563–78.

Jarosińska, Elżbieta and Katarzyna Pierzga (2015). Estimating flood quantiles on the basis of multi-event rainfall simulation – case study. *Acta Geophysica*. Available at: https://doi.org/10.1515/acgeo-2015-0046.

Jenkins, Geoff, Matthew Perry, John Prior, and (as contributor) Phil Woodworth (2009). *UKCIP08: The Climate of the United Kingdom and Recent Trends*. Exeter: Met Office Hadley Centre.

Jevrejeva, Svetlana, L. P. Jackson, Aslak Grinsted, Daniel Lincke, and Ben Marzeion (2018). Flood damage costs under the sea level rise with warming of 1.5 C and 2 C. *Environmental Research Letters*, 13(7): 074014.

Jin, Zhenong, Qianlai Zhuang, Jiali Wang, et al. (2017). The combined and separate impacts of climate extremes on the current and future US rainfed maize and soybean production under elevated CO_2. *Global Change Biology*, 23(7): 2687–704.

Johnson, F. and Sharma, A. (2012). A nesting model for bias correction of variability at multiple time scales in general circulation model precipitation simulations. *Water Resources Research*, 48(1): 10.1029/2011WR010464

Jun, M., R. Knutti, and D. W. Nychka (2008). Local eigenvalue analysis of CMIP3 climate model errors. *Tellus A*, 60: 992–1000. Available at: http://dx.doi.org/10.1111/j.1600-0870.2008.00356.x

Jungclaus, J. H., Noel Keenlyside, M. Botzet, et al. (2006). Ocean circulation and tropical variability in the coupled model ECHAM5/MPI-OM. *Journal of Climate*, 19(16): 3952–72.

Kanamitsu, Masao and Laurel DeHaan (2011). The Added Value Index: A new metric to quantify the added value of regional models. *Journal of Geophysical Research: Atmospheres*, 116(D11). doi: 116, D11106, doi:10.1029/2011JD015597.

Kannan, S. and S. Ghosh (2013). A nonparametric kernel regression model for downscaling multisite daily precipitation in the Mahanadi basin, *Water Resources Research*, 49: 1360–85. doi:10.1002/wrcr.20118

Kaptué, Armel T., Niall P. Hanan, Lara Prihodko, and Jorge A. Ramirez (2015). Spatial and temporal characteristics of rainfall in Africa: Summary statistics for temporal downscaling. *Water* Resources *Research*, 51(4): 2668–79.

Karl, Thomas R., Wei-Chyung Wang, Michael E. Schlesinger, Richard W. Knight, and David Portman (1990). A method of relating general circulation model simulated climate to the observed local climate. Part I: Seasonal statistics. *Journal of Climate*, 3(10): 1053–79.

Kay, Jennifer E., Clara Deser, A. Phillips, et al. (2015). The Community Earth System Model (CESM) large ensemble project: A community resource for studying climate change in the presence of internal climate variability. *Bulletin of the American Meteorological Society*, 96(8): 1333–49.

Kendon, Elizabeth J., Nigel M. Roberts, Hayley J. Fowler, et al. (2014). Heavier summer downpours with climate change revealed by weather forecast resolution model. *Nature Climate Change,* 4(June): 570.

Kim, D., H. Cho, C. Ono, and M. Choi (2017a). Let-It-Rain: A web application for stochastic point rainfall generation at ungaged basins and its applicability in runoff and flood modeling. *Stochastic Environmental Research and Risk Assessment,* 31(4): 1023–43.

Kim, Jang-Gyeong, Hyun-Han Kwon, and Dongkyun Kim (2017b). A hierarchical Bayesian approach to the modified Bartlett-Lewis Rectangular Pulse Model for a joint estimation of model parameters across stations. *Journal of Hydrology*, 544(January): 210–23.

Kling, G. W., K. Hayhoe, L. B. Johnson, J. J. et al. (2003). *Confronting Climate Change in the Great Lakes Region: Impacts on Our Communities and Ecosystems*. Washington, DC: Ecological Society of America.

Knox, Jerry, Tim Hess, Andre Daccache, and Tim Wheeler (2012). Climate change impacts on crop productivity in Africa and South Asia. *Environmental Research Letters: ERL,* 7(3): 034032.

Knutson, T. R., R. Zhang, and L. W. Horowitz (2016). Prospects for a prolonged slowdown in global warming in the early 21st century. *Nature Communcations*, 7: 13676. Available at: http://dx.doi.org/10.1038/ncomms13676

Knutti, R. and G. C. Hegerl (2008). The equilibrium sensitivity of the Earth's temperature to radiation changes. *Nature Geoscience,* 1(11): 735.

Knutti, R., D. Masson, and A. Gettelman (2013). Climate model genealogy: Generation CMIP5 and how we got there. *Geophysical Research Letters*, 40: 1194–9. Available at: http://dx.doi.org/10.1002/grl.50256

Knutti, R., J. Rogelj, J. Sedláček, and E. M. Fischer (2016). A scientific critique of the two-degree climate change target. *Nature Geoscience*, 9: 13–18. Available at: http://dx .doi.org/10.1038/ngeo2595

Knutti, R. and J. Sedláček (2013). Robustness and uncertainties in the new CMIP5 climate model projections. *Nature Climate Change*, 3: 369–73. Available at: http://dx.doi .org/10.1038/nclimate1716.

Knutti, R., J. Sedláček, B. M. Sanderson, et al. (2017). A climate model projection weighting scheme accounting for performance and interdependence. *Geophysical Research Letters*, 44: 1909–18. Available at: http://dx.doi.org/10.1002/2016GL072012

Kompas, Tom, Van Ha Pham, and Tuong Nhu Che (2018). The effects of climate change on GDP by country and the global economic gains from complying with the Paris Climate Accord. *Earth's Future,* 6(8): 1153–73.

Kopp, R. E., K. Hayhoe, D. R. Easterling, et al. (2017). Potential surprises – compound extremes and tipping elements. In D. J. Wuebbles, D. W. Fahey, K. A. Hibbard, et al., (eds.), *Climate Science Special Report: Fourth National Climate Assessment, Volume I*. Washington, DC.: U.S. Global Change Research Program: 411–29. doi: 10.7930/ J0GB227J

Kossieris, Panagiotis, Christos Makropoulos, Christian Onof, and Demetris Koutsoyiannis (2018). A rainfall disaggregation scheme for sub-hourly time scales: Coupling a Bartlett-Lewis based model with adjusting procedures. *Journal of Hydrology*, 56(January): 980–92.

Kotamarthi, Rao, Linda Mearns, Katharine Hayhoe, Christoper L. Castro, and Donald Wuebbles (2016). *Use of Climate Information for Decision-Making and Impacts*

Research: State of Our Understanding. Argonne, WI: Argonne National Laboratory. Available at: https://apps.dtic.mil/docs/citations/AD1029525

Kotlarski, S., K. Keuler, and O. B. Christensen (2014). Regional climate modeling on European scales: A joint standard evaluation of the EURO-CORDEX RCM Ensemble. *Geoscientific Model.* Available at: https://pure.mpg.de/rest/items/item_2060613/component/file_2060614/content

Krasnopolsky, Vladimir M., Michael S. Fox-Rabinovitz, and Alexei A. Belochitski (2013). Using ensemble of neural networks to learn stochastic convection parameterizations for climate and numerical weather prediction models from data simulated by a cloud resolving model. *Advances in Artificial Neural Systems.* Available at: https://doi.org/10.1155/2013/485913

Kreienkamp, Frank, Heike Huebener, Carsten Linke, and Arne Spekat (2012). Good practice for the usage of climate model simulation results – a discussion paper. *Environmental Systems Research* 1(1): 9.

Larson, V. E. and J. Golaz (2005). Using probability density functions to derive consistent closure relationships among higher-order moments. *Monthly Weather Review,* 133:1023–42. Available at: https://doi.org/10.1175/MWR2902.1

Lazante, J. R., K. W. Dixon, M. J. Nath, C. E. Whitlock, and D. Adams-Smith (2018). Some pitfalls in statistical downscaling of future climate. *Bulletin of the American Meteorological Society.* DOI:10.1175/BAMS-D-17-0046.1

Lee, M.-H., M. Lu, E.-S. Im, D.-H. Bae (2019). Added value of dynamical downscaling for hydrological projections in the Chungju Basin, Korea. *International Journal of Climatology,* 39: 516–31. Available at: https://doi.org/10.1002/joc.5825

Lemos, Maria Carmen, Christine J. Kirchhoff, and Vijay Ramprasad (2012). Narrowing the climate information usability gap. *Nature Climate Change.* Available at: https://doi.org/10.1038/nclimate1614

Lemos, Maria Carmen and Richard B. Rood (2010). Climate projections and their impact on policy and practice. *Wiley Interdisciplinary Reviews: Climate Change,* 1(5): 670–82.

Lempert, Robert J., Steven W. Popper, and Steven C. Bankes (2010). Robust decision making: Coping with uncertainty. *The Futurist,* 44(1): 47.

Leung, L. R., T. Ringler, W. D. Collins, M. Taylor, and M. Ashfaq (2013). A hierarchical evaluation of regional climate simulations. *Eos Trans. AGU,* 94(34): 297

Li, H., M. Kanamitsu, S.-Y. Hong, et al. (2014). Projected climate change scenario over California by a regional ocean–atmosphere coupled model system. *Climatic Change,* 122(4): 609–19.

Li, R., S. Lv, B. Han, Y. Gao, and X. Meng (2017). Projections of South Asian summer monsoon precipitation based on 12 CMIP5 Models. *International Journal of Climatology,* 37(1): 94–108.

Li, S., P. W. Mote, D. E. Rupp, et al. (2015). Evaluation of a regional climate modeling effort for the Western United States using a Superensemble from Weather@ Home. *Journal of Climate,* 28(19): 7470–88.

Li Liu, D. and Zuo, H. (2012). Statistical downscaling of daily climate variables for climate change impact assessment over New South Wales, Australia. *Climatic Change, 115* (3–4): 629–66.

Liang, X.-Z., M. Xu, X. Yuan, et al. (2012). Regional climate–weather research and forecasting model. *Bulletin of the American Meteorological Society,* 93(9): 1363–87.

Lichter, M., A. Vafeidis, R. Nicholls, and G. Kaiser (2011). Exploring data-related uncertainties in analyses of land area and population in the "Low-Elevation Coastal Zone" (LECZ). *Journal of Coastal Research,* 27(4): 757–68.

Lipscomb, W. H., J. G. Fyke, M. Vizcaíno, et al. (2013). Implementation and initial evaluation of the Glimmer Community ice sheet model in the Community Earth System Model. *Journal of Climate*, 26: 7352–71. Available at: https://doi.org/10.1175/JCLI-D-12-00557.1

Liston, Glen E. and Kelly Elder (2006). A meteorological distribution system for high-resolution terrestrial modeling (MicroMet). *Journal of Hydrometeorology*. Available at: https://doi.org/10.1175/jhm486.1

Liu, Y., J. Stanturf, and S. Goodrick (2010). Trends in global wildfire potential in a changing climate. *Forest Ecology and Management*, 259(4): 685–97. Available at: https://doi.org/10.1016/j.foreco.2009.09.00

Lo, J. C.-F., Z.-L. Yang, and R. A. Pielke Sr. (2008). Assessment of three dynamical climate downscaling methods using the Weather Research and Forecasting (WRF) Model. *Journal of Geophysical Research*, 113(D9): 1306.

Lombardo, F., E. Volpi, D. Koutsoyiannis, and F. Serinaldi (2017). A theoretically consistent stochastic cascade for temporal disaggregation of intermittent rainfall. *Water Resources Research*, 53(6): 4586–605.

Lorenz, Susanne, Suraje Dessai, Piers M. Forster, and Jouni Paavola (2017). Adaptation planning and the use of climate change projections in local government in England and Germany. *Regional Environmental Change,* 17(2): 425–35.

Lu, J., G. J. Carbone, and J. M. Grego (2019). Uncertainty and hotspots in 21st century projections of agricultural drought from CMIP5 models. *Nature, Scientific Reports,* 9: 4922. Available at: https://doi.org/10.1038/s41598–019-41196-z

Lukas, J., J. Barsugli, N. Doesken, I. Rangwala, and K. Wolter (2014). Climate Change in Colorado: A Synthesis to Support Water Resources Management and Adaptation. A report for the Colorado Water Conservation Board by the Western Water Assessment. 114 pp.

Lv, Z., Y. Zhu, X., Liu, et al. (2018). Climate change impacts on regional rice production in China. *Climatic Change*, 147: 523–37

Mabuchi, K., Y. Sato, and H. Kida (2002). Verification of the climatic features of a regional climate model with BAIM. *Journal of the Meteorological Society of Japan*. Available at: https://doi.org/10.2151/jmsj.80.621

Magrin, G. O., J. A. Marengo, J.-P. Boulanger, et al. (2014). Central and South America. In: V. R. Barros, C. B. Field, D. J. Dokken, et al. (eds.), *Climate Change 2014: Impacts, Adaptation, and Vulnerability. Part B: Regional Aspects. Contribution of Working Group II to the Fifth Assessment Report of the Intergovernmental Panel on Climate Change*. Cambridge, UK and New York: Cambridge University Press: 1499–556.

Manabe, S. and K. Bryan (1969). Climate calculations with a combined ocean-atmosphere model. *Journal of the Atmospheric Science*, 26: 786–9.

Manabe, S. and R. Wetherald (1975). The effects of doubling with CO_2 concentration on the climate of a general circulation model. *Journal of the Atmospheric Science*, 32: 3–15,

Manzanas, R., J. M. Gutiérrez, J. Fernández, et al. (2018). Dynamical and statistical downscaling of seasonal temperature forecasts in Europe: Added value for user applications. *Climate Services,* 9: 44–56.

Maraun, D. and M. Widmann (2018). *Statistical Downscaling and Bias Correction for Climate Research*. Cambridge: Cambridge University Press: 347 pp.

Maurer, E. P. and Hidalgo, H. G. (2008). Utility of daily vs. monthly large-scale climate data: an intercomparison of two statistical downscaling methods, *Hydrology and Earth System Science*, 12: 551–63. Available at: https://doi.org/10.5194/hess-12-551-2008

Maurer, J. M., J. M. Schaefer, S. Rupper, and A. Corley (2019). Acceleration of ice loss across the Himalayas over the past 40 years. *Science Advances*, 5(6): eaav7266.

Maurer, E. P., A. W. Wood, J. C. Adam, D. P. Lettenmaier, and B. Nijssen (2002). A long-term hydrologically based dataset of land surface fluxes and states for the conterminous United States. *Journal of Climate*, 15(22): 3237–51.

McFarlane, N. (2011). Parameterizations: Representing key processes in climate models without resolving them. *Wiley Interdisciplinary Reviews: Climate Change*, 2(4): 482–97.

McGinnis, S., D. Nychka, and L. O. Mearns (2015). A new distribution mapping technique for climate model bias correction. In V. Lakshmanan, E. Gilleland, A. McGovern, M. Tingley, (eds.), *Machine Learning and Data Mining Approaches to Climate Science: Proceedings of the Fourth International Workshop on Climate Informatics*. Cham, Switzerland: Springer: 91–99. doi:10.1007/978-3-319-17220-0

McGranahan, Gordon, Deborah Balk, and Bridget Anderson (2007). The rising tide: Assessing the risks of climate change and human settlements in low elevation coastal zones. *Environment and Urbanization,* 19(1): 17–37.

Md Hafijur Rahaman Khan, A. Rahman, C. Luo, S. Kumar, G. M. A. and Islam, M. A. Hossain (2019). Detection of changes and trends in climatic variables in Bangladesh during 1988–2017, *Heliyon*, 5(3). doi:10.1016/j.heliyon.2019.e01268

Mearns, L. O., R. Arritt, S. Biner, et al. (2012). The North American Regional Climate Change Assessment Program: Overview of Phase I results. *Bulletin of the American Meteorological Society*, 93: 1337–62. Available at: https://doi.org/10.1175/BAMS-D-11-00223.1

Mearns, L. O., I. Bogardi, F. Giorgi, I. Matyasovszky, and M. Palecki (1999). Comparison of climate change scenarios generated from regional climate model experiments and statistical downscaling. *Journal of Geophysical Research*, 104(D6): 6603–21.

Mearns, Linda O., Melissa S. Bukovsky, Sarah C. Pryor, and Victor Magaña (2014). Downscaling of climate information. In George Ohring (ed.), *Climate Change in North America*. Cham: Springer International Publishing: 201–50.

Mearns, Linda O., Melissa S. Bukovsky, and Vanessa J. Schweizer (2017). Potential value of expert elicitation for determining differential credibility of regional climate change simulations: An exercise with the NARCCAP co-PIs for the Southwest Monsoon region of North America. *Bulletin of the American Meteorological Society*, 98 (1): 29–35.

Mearns, L. O., Easterling, W., Hays, C., and Marx, D. (2001). Comparison of agricultural impacts of climate change calculated from high and low resolution climate change scenarios: Part I. The uncertainty due to spatial scale. *Climatic Change*, 51(2): 131–72.

Mearns, L. O., F. Giorgi, P. Whetton, et al. (2003). Guidelines for Use of Climate Scenarios Developed from Regional Climate Model Experiments. Available at: http://climate-action.engin.umich.edu/downscaling/Mearns_IPCC_Downscaling_Guidelines_IPCC_2004.pdf

Mearns, L. O., D. P. Lettenmaier, and S. McGinnis (2015). Uses of results of regional climate model experiments for impacts and adaptation studies: The example of NARCCAP. *Current Climate Change Reports*, 1(1): 1–9.

Mearns, L. O., S. Sain, L. R. Leung, et al. (2013). Climate change projections of the North American Regional Climate Change Assessment Program (NARCCAP). *Climatic Change*, 120(4): 965–75.

Meehl, G. A., C. Covey, T. Delworth, et al. (2007). The WCRP CMIP3 multimodel dataset: A new era in climate model research. *Bulletin of the American Meteorological Society*, 88: 1383–94.

Meehl, G. A., A. Hu, B. D. Santer, and S.-P. Xie (2016). Contribution of the Interdecadal Pacific Oscillation to twentieth-century global surface temperature trends. *Nature Climate Change*, 6: 1005–8. Available at: http://dx.doi.org/10.1038/nclimate3107

Michelangeli, P.-A., M. Vrac, and H. Loukos (2009). Probabilistic downscaling approaches: application to wind cumulative distribution functions. *Geophysical Research Letters*, 36(11): 136.

Miguez-Macho, G., G. L. Stenchikov, and A. Roboc. (2005). Regional climate simulations over North America: Interaction of local processes with improved large-scale flow. *Journal of Climate*, 18(8): 1227–46.

Ministry of Environment, Ministry of Agriculture, Forestry and Fisheries (2018). Climate change in Japan and its impacts. Available at: www.env.go.jp/earth/tekiou/pamph2018_full_Eng.pdf

Monerie, P., B. Fontaine, and P. Roucou (2012). Expected future changes in the African Monsoon between 2030 and 2070 using some CMIP3 and CMIP5 models under a medium-low RCP RCPScenario. *Journal of Geophysical Research, D: Atmospheres*, 117 (D16). Available at: https://onlinelibrary.wiley.com/doi/pdf/10.1029/2012JD017510

Morrison, H. and A. Gettelman (2008). A new two-moment bulk stratiform cloud microphysics scheme in the community atmosphere model, Version 3 (cam3). Part I: Description and numerical tests. *Journal of Climate*, 21: 3642–59. Available at: https://doi.org/10.1175/2008JCLI2105.1

Moser, S. C., M. A. Davidson, P. Kirshen, et al. (2014). Ch. 25: Coastal zone development and ecosystems. In J. M. Melillo, Terese (T. C.) Richmond, and G. W. Yohe, (eds.), *Climate Change Impacts in the United States: The Third National Climate Assessment*, U.S. Global Change Research Program: 579–618. doi:10.7930/J0MS3QNW

Moss, R. H., J. A. Edmonds, K. A. Hibbard, et al. (2010). The next generation of scenarios for climate change research and assessment. *Nature*, 463: 747–56. http://dx.doi.org/10.1038/nature08823

Mukherjee, S., S. Aadhar, D. Stone, and V. Mishra (2018). Increase in extreme precipitation events under anthropogenic warming in India. *Weather and Climate Extremes*, 20(June): 45–53.

Müller, H. and U. Haberlandt (2018). Temporal rainfall disaggregation using a multiplicative cascade model for spatial application in urban hydrology. *Journal of Hydrology*, 556(January): 847–64.

Muluye, Getnet Y. (2011). Implications of medium-range numerical weather model output in hydrologic applications: Assessment of skill and economic value. *Journal of Hydrology*, 400(3–4): 448–64.

Murakami, H., B. Wang, and A. Kitoh (2011). Future change of Western North Pacific typhoons: Projections by a 20-km-mesh global atmospheric model. *Journal of Climate*, 24: 1154–69. Available at: https://doi.org/10.1175/2010JCLI3723.1

Murphy, J. M., B. B. B. Booth, M. Collins, et al. (2007). A methodology for probabilistic predictions of regional climate change from perturbed physics ensembles. *Philosophical Transactions. Series A, Mathematical, Physical, and Engineering Sciences*, 365 (1857): 1993–2028.

Murphy J. M., G. R. Harris, D. M. H. Sexton, et al. (2018). UKCP18 Land Projections: Science Report. Bracknell: Met Office. Available at: www.metoffice.gov.uk/pub/data/weather/uk/ukcp18/science-reports/UKCP18-Land-report.pdf

Myhre, G., E. Highwood, K. Shine, and F. Stordal (1998). New estimates of radiative forcing due to well mixed greenhouse gases. *Geophysical Research Letters*, 25(14): 2715–18. doi:10.1029/98GL01908

Nakicenovic, N., J. Alcamo, G. Davis, et al. (2000). *IPCC Special Report on Emissions Scenarios*. N. Nakicenovic and R. Swart (eds.). Cambridge: Cambridge University Press, 2000. Available at: www.ipcc.ch/ipccreports/sres/emission/index.php?idp=0

NAS, 2012: *A National Strategy for Advancing Climate Modeling*. Washington, DC.: National Academy Press.

National Oceanic and Atmospheric Administration (NOAA) (2013). National Coastal Population Report. A product of the NOAA State of the Coast Report Series, a publication of the National Oceanic and Atmospheric Administration, Department of Commerce, developed in partnership with the U.S. Census Bureau. Available at: http://stateofthecoast.noaa.gov

Niang, I., O. C. Ruppel, M. A. Abdrabo, et al. (2014). Africa. In V. R. Barros, C. B., Field, D. J. Dokken, et al. (eds.), *Climate Change 2014: Impacts, Adaptation, and Vulnerability. Part B: Regional Aspects. Contribution of Working Group II to the Fifth Assessment Report of the Intergovernmental Panel on Climate Change*. Cambridge, UK and New York: Cambridge University Press: 1199–265.

Nicholls, R. J., S. E. Hanson, J. A. Lowe, et al. (2014). Sea-level scenarios for evaluating coastal impacts. *Wiley Interdisciplinary Reviews. Climate Change*, 5(1): 129–50.

Nowicki, S. M. J., T. Payne, E. Larour, et al. (2016). Ice sheet model intercomparison project (ISMIP6) contribution to CMIP6. *Geoscientific Model Development,* 9(12): 4521.

NYC Mayors' Office of Recovery and Resiliency (2019). Climate Resiliency Design Guidelines. Available at: www1.nyc.gov/assets/orr/pdf/NYC_Climate_Resiliency_Design_Guidelines_v3-0.pdf

NYCPCC, New York City Panel on Climate Change (2019). Report Executive Summary 2019. *Annals of the New York Academy of Sciences*, 1439: 11–21. doi:10.1111/nyas.14008

O'Neill, Brian C., Elmar Kriegler, Keywan Riahi, et al. (2014). A new scenario framework for climate change research: The concept of shared socioeconomic pathways. *Climatic Change*, 122(3): 387–400.

Oh, S.-G., J.-H. Park, S.-H. Lee, and M.-S. Suh (2014). Assessment of the RegCM4 over East Asia and future precipitation change adapted to the RCP Scenarios. *Journal of Geophysical Research, D: Atmospheres*, 119(6): 2913–27.

Pachauri, R. K., M. R. Allen, V. R. Barros, et al. (2014). *Climate Change 2014: Synthesis Report. Contribution of Working Groups I, II and III to the Fifth Assessment Report of the Intergovernmental Panel on Climate Change*, (eds.) R. K. Pachauri and L. Meyer. Geneva, Switzerland: IPCC.

Paeth, H. and B. Manning (2013). On the added value of regional climate modeling in climate change Assessment. *Climate Dynamics*, 41: 1057–66.

Pal, J. S., F. Giorgi, X. Bi, et al. (2007). Regional climate modeling for the developing world: The ICTP RegCM3 and RegCNET. *Bulletin of the American Meteorological Society*, 88(9): 1395–410.

Palomino-Lemus, Re., S. Córdoba-Machado, S. R. Gámiz-Fortis, Y. Castro-Díez, and M. J. Esteban-Parra (2018). High-resolution boreal winter precipitation projections over tropical America from CMIP5 models. *Climate Dynamics*, 51(5–6): 1773–92.

Parmesan, C. and G. Yohe, (2003). A globally coherent fingerprint of climate change impacts across natural systems, *Nature*, 421(6918): 37–42.

Parry, M. and T. Carter (1996). *Climate Impact and Adaptation Assessment*. New York: Earthscan.

PBMC (2013). Executive summary: Scientific basis of climate change – Contribution from Grupo de Trabalho 1 (GT1, acronym for the Working Group 1) to the Primeiro Relatório de Avaliação Nacional sobre Mudanças Climáticas (RAN1) of the Painel Brasileiro de Mudanças Climáticas (PBMC), (eds.) T. Ambrizzi, M. Araujo. Rio de Janeiro: Universidade Federal do Rio de Janeiro: 24 pp.

Perry, W. J., and J. P. Abizaid (2014). Ensuring a strong US Defense for the future: The National Defense Panel Review of the 2014 Quadrennial Defense Review. UNITED

STATES INST OF PEACE WASHINGTON DC. Available at: https://apps.dtic.mil/docs/citations/ADA604218.

Peters, G. P., R. M. Andrew, T. Boden, et al. (2013). The challenge to keep global warming below 2 °C. *Nature Climate Change*, 3(4–6). Available at: https://doi.org/10.1038/nclimate1783.

Piao, S. L. et al. (2010). The impacts of climate change on water resources and agriculture in China. *Nature*, 467: 43–51

Pierce, D. W., D. R. Cayan, and B. L. Thrasher (2014). Statistical downscaling using Localized Constructed Analogs (LOCA). *Journal of Hydrometeorology,* 15(6): 2558–85.

Pinto, J. G., C. P. Neuhaus, and G. C. Leckebusch (2010). Estimation of Wind Storm Impacts over Western Germany under Future Climate Conditions Using a Statistical–Dynamical Downscaling Approach. Tellus A: Dynamic. Available at: www.tandfonline.com/doi/abs/10.1111/j.1600-0870.2009.00424.x.

Prein, Andreas F., Wolfgang Langhans, Giorgia Fosser et al. (2015). A review on regional convection permitting climate modeling: Demonstrations, prospects, and challenges. *Reviews of Geophysics*, 53: 323–61.

Prudhomme, Christel and Helen Davies (2009). Assessing uncertainties in climate change impact analyses on the river flow regimes in the UK. Part 2: Future climate. *Climatic Change*, 93(1–2): 177–95. Available at: https://doi.org/10.1007/s10584–008-9461-6.

Ramanathan, V. and J. R. Coakley, Jr. (1978). Climate modeling through radiative-convective models. *Reviews of Geophysics and Space Physics*, 16: 465–89,

Rhoades, Alan M., Xingying Huang, Paul A. Ullrich, and Colin M. Zarzycki (2016). Characterizing Sierra Nevada snowpack using variable-resolution CESM. *Journal of Applied Meteorology and Climatology,* 55(1): 173–96. Available at: https://doi.org/10.1175/jamc-d-15-0156.1.

Richardson, M., K. Cowtan, E. Hawkins, and M. B. Stolpe (2016). Reconciled climate response estimates from climate models and the energy budget of Earth. *Nature Climate Change*, 6: 931–5. Available at: http://dx.doi.org/10.1038/nclimate3066.

Rockel, B., A. Will, and A. Hense (2008). The Regional Climate Model COSMO-CLM (CCLM). *Meteorologische Zeitschrift*. Available at: https://doi.org/10.1127/0941-2948/2008/0309.

Rössler, O., A. M. Fischer, D. Maraun et al. (2017). Challenges to link climate change data provision and user needs – perspective from the COST-action VALUE. *International Journal of Climatology*, 50: 7541

Rummukainen, M. (2016). Added value in regional climate modeling. *Wires Climate Change*, 7: 145–59.

Ruth, Matthias and Dana Coelho (2007). Understanding and Managing the Complexity of Urban Systems under Climate Change. *Climate Policy*. Available at: https://doi.org/10.3763/cpol.2007.0716.

Ryu, J.-H. and K. Hayhoe (2013). Understanding the sources of Caribbean precipitation biases in CMIP3 and CMIP5 simulations. *Climate Dynamics*, 42: 3233–52. Available at: http://dx.doi.org/10.1007/s00382–013-1801-1

Sachindra, D. A., F. Huang, A. Barton, and B. J. C. Perera (2014). Statistical downscaling of General Circulation Model outputs to precipitation – Part 2: Bias-correction and future projections. *International Journal of Climatology*, 34(11): 3282–303.

Salvi, K., S. Ghosh, and A. Ganguly (2016). Credibility of statistical downscaling under nonstationary climate. *Climate Dynamics*, 46(5): 1991–2023. doi: 10.1007/s00382-015-2688-9

Salvi, K., S. Kannan, and S. Ghosh (2013). High-resolution multisite daily rainfall projections in India with statistical downscaling for climate change impacts assessment, *Journal of Geophysical Research: Atmospheres*, 118: 3557–78. doi:10.1002/jgrd.50280.

Sanderson, B. M. (2011). A multimodel study of parametric uncertainty in predictions of climate response to rising greenhouse gas concentrations. *Journal of Climate*, 24(5): 1362–77.

Sanderson, B. M., R. Knutti, and P. Caldwell (2015). A representative democracy to reduce interdependency in a multimodel ensemble. *Journal of Climate*, 28: 5171–94. Available at: http://dx.doi.org/10.1175/JCLI-D-14-00362.1

Sanderson, B. M., M. Wehner, and R. Knutti (2017). Skill and independence weighting for multi-model assessments. *Geoscientific Model Development*, 10: 2379–95.

Santer, B. D., S. Solomon, G. Pallotta, et al. (2017). Comparing tropospheric warming in climate models and satellite data. *Journal of Climate*, 30: 373–92. Available at: http://dx.doi.org/10.1175/JCLI-D-16-0333.1

Schellnhuber, H. J., S. Rahmstorf, and R. Winkelmann (2016). Why the right climate target was agreed in Paris. *Nature Climate Change*, 6, 649–53. Available at: http://dx.doi.org/ 10.1038/nclimate301

Scher, S. (2018). Toward data-driven weather and climate forecasting: Approximating a simple general circulation model with deep learning. *Geophysical Research Letters*. Available at: https://doi.org/10.1029/2018gl080704

Schlesinger, M. and J. F. B. Mitchell (1985). Model projections of the equilibrium climatic response to increases carbon dioxide: The potential climatic effects of increasing carbon dioxide, *Rep. DOE/ER-0237*: 83–147.

Schlesinger M. and J. F. B. Mitchell (1986). *Model Projections of the Equilibrium Climatic Response to Increased Carbon Dioxide*. No. UCRL-15807. Corvallis, OR: Oregon State Univ; Bracknell: Meteorological Office (UK).

Schneider, Tapio, Shiwei Lan, Andrew Stuart, and João Teixeira (2017). Earth system modeling 2.0: A blueprint for models that learn from observations and targeted high-resolution simulations. *Geophysical Research Letters*, 44(24): 12–396.

Sellers, W. D. (1969). A global climatic model based on the energy balance of the Earth atmosphere system. *Journal of Applied Meteorology*, 21: 391–400.

Seo, K.-H., J. Ok, J.-H. Son, and D.-H. Cha (2013). Assessing future changes in the East Asian summer monsoon using CMIP5 coupled models. *Journal of Climate,* 26 (19): 7662–75.

Shao, Q., L. Zhang, and Q. J. Wang (2016). A hybrid stochastic-weather-generation method for temporal disaggregation of precipitation with consideration of seasonality and within-month variations. *Stochastic Environmental Research and Risk Assessment*, 30(6): 1705–24.

Sheffield, Justin, Gopi Goteti, and Eric F. Wood (2006). Development of a 50-year high-resolution global dataset of meteorological forcings for land surface modeling. *Journal of Climate*, 19(13): 3088–111.

Sheffield, J. and E. F. Wood (2007). Characteristics of global and regional drought, 1950–2000: Analysis of soil moisture data from off-line simulation of the terrestrial hydrologic cycle. *Journal of Geophysical Research, D: Atmospheres*, 112 (D17). Available at: https://agupubs.onlinelibrary.wiley.com/doi/abs/10.1029/2006JD008288

Shtiliyanova, Anastasiya, Gianni Bellocchi, David Borras, et al. (2017). Kriging-based approach to predict missing air temperature data. *Computers and Electronics in Agriculture*, 142(November): 440–9.

Sikorska, A. E. and J. Seibert (2018). Appropriate temporal resolution of precipitation data for discharge modelling in pre-alpine catchments. *Hydrological Sciences Journal*. Available at: www.tandfonline.com/doi/abs/10.1080/02626667.2017.1410279

Skamarock, W., J. B. Klemp, J. Dudhia, et al. (2008). A description of the Advanced Research WRF Version 3. NCAR Technical Note, NCAR/ TN-475+STR, 123 pp

Skelton M., J. J. Porter, S. Dessai, et al. (2017). The social and scientific values that shape national climate scenarios: A comparison of the Netherlands, Switzerland and the UK. *Regional Environmental Change*, 17(8): 2325–38.

Skourkeas, Anastasios, Fotini Kolyva-Machera, and Panagiotis Maheras (2013). Improved statistical downscaling models based on canonical correlation analysis, for generating temperature scenarios over Greece. *Environmental and Ecological Statistics*. Available at: https://doi.org/10.1007/s10651-012-0228-x

Smith, M. J., Palmer, P. I., Purves, et al. (2014). Changing how earth system modeling is done to provide more useful information for decision making, science, and society. *Bulletin of the American Meteorological Society*, 95: 1453–64. doi:10.1175/BAMS-D-13-00080.1

Smith, T. and B. Bookhagen (2018). changes in seasonal snow water equivalent distribution in high mountain Asia (1987 to 2009). *Science Advances,* 4(1): e1701550.

SNAP (Scenarios Network for Alaska and Arctic Planning), (2016). *About SNAP Data*, University of Alaska. Available at www.snap.uaf.edu/tools/data-downloads

So, Byung-Jin, Jin-Young Kim, Hyun-Han Kwon, and Carlos H. R. Lima (2017). Stochastic extreme downscaling model for an assessment of changes in rainfall intensity-duration-frequency curves over South Korea using multiple regional climate models. *Journal of Hydrology,* 553(October): 321–37.

Soares, P. M. M., Maraun, D, Brands, S. et al. (2019). Process-based evaluation of the VALUE perfect predictor experiment of statistical downscaling methods. *International Journal of Climatology*. 39: 3868–93. Available at https://doi.org/10.1002/joc.5911

Solomon, S., Qin, D., Manning, M., et al. (eds.) (2007). *Climate Change 2007 – The Physical Science Basis: Working Group I Contribution to the Fourth Assessment Report of the IPCC* (Vol. 4). Cambridge: Cambridge University Press.

Spak, Scott, Tracey Holloway, Barry Lynn, and Richard Goldberg (2007). A comparison of statistical and dynamical downscaling for surface temperature in North America. *Journal of Geophysical Research,* 112(D8): 1645.

Stocker, Thomas F., Dahe Qin, Gian-Kasper Plattner, et al. (2013). Climate phenomena and their relevance for future regional climate change. In *Climate Change 2013: The Physical Science Basis. Contribution of Working Group I to the Fifth Assessment of the Intergovernmental Panel on Climate Change*, (eds.) Thomas F. Stocker, Dahe Qin, Gian-Kasper Plattner et al. 1217–308. Cambridge: Cambridge University Press.

Stoner, Anne M. K., Katharine Hayhoe, Xiaohui Yang, and Donald J. Wuebbles (2013). An asynchronous regional regression model for statistical downscaling of daily climate variables. *International Journal of Climatology*, 33(11): 2473–94.

Stoner, A., Hayhoe, K. Dixon, J. Lanzante, and I. Scott-Fleming (2017). *Comparing the Performance of Multiple Statistical Downscaling Approaches Using a Perfect Model Framework*. Presented at the Annual Meeting of the American Meteorological Society. Seattle WA.

Subyani, Ali M. and Nassir S. Al-Amri (2015). IDF curves and daily rainfall generation for Al-Madinah City, Western Saudi Arabia. *Arabian Journal of Geosciences*, 8(12): 11107–19.

Sun, F., N. Berg, A. Hall, M. Schwartz, and D. Walton (2019). Understanding end-of-century snowpack changes over California's Sierra Nevada. *Geophysical Research Letters*, 46: 933–43. Available at: https://doi.org/10.1029/2018GL080362

Sun, L., K. E. Kunkel, L. E. Stevens, et al. (2015b). *Regional Surface Climate Conditions in CMIP3 and CMIP5 for the United States: Differences, Similarities, and Implications for the U.S. National Climate Assessment*. National Oceanic and Atmospheric Administration, National Environmental Satellite, Data, and Information Service, 111 pp. Available at: http://dx.doi.org/10.7289/V5RB72KG.

Sun, F., D. B. Walton, and A. Hall, (2015a). A hybrid dynamical-statistical technique: Part II: End of century warming projections predict a new climate state in the Los Angeles region. *Journal of Climate*, 28: 4618–36

Suppiah, R., M. Collier, S. Jeffrey, L. et al. (2013). Simulated and projected summer rainfall in tropical Australia: Links to atmospheric circulation using the CSIRO-Mk3.6 climate model. *Australian Meteorological and Oceanographic Journal*, 63(1): 15–26.

Swaminathan, Ranjini, Mohan Sridharan, and Katharine Hayhoe (2018). A computational framework for modelling and analyzing ice storms. *arXiv preprint arXiv:1805.04907*.

Swart, R. J., K. de Bruin, S. Dhenain et al. (2017). Developing climate information portals with users: Promises and pitfalls. *Climate Services*, 6: 12–22.

Sweet, W. V., R. E. Kopp, C. P. Weaver, et al. (2017). *Global and Regional Sea Level Rise Scenarios for the United States*. NOAA Technical Report NOS CO-OPS 083. NOAA/NOS Center for Operational Oceanographic Products and Services.

Syafrina, A. H., A. Norzaida, and O. Noor Shazwani (2018). Stochastic modeling of rainfall series in Kelantan using an advanced weather generator. *ETASR*, 8: 2537–41.

Tang, Jianping, Xiaorui Niu, Shuyu Wang et al. (2016). Statistical downscaling and dynamical downscaling of regional climate in China: Present climate evaluations and future climate projections. *Journal of Geophysical Research, D: Atmospheres*, 121(5): 2110–29.

Taylor, K. E. (2001). Summarizing multiple aspects of model performance in a single diagram. *Journal of Geophysical Research, WMO TD-732*, 106(D7): 7183–92.

Taylor, K. E., R. J. Stouffer, and G. A. Meehl (2012). An overview of CMIP5 and the experiment design. *Bulletin of the American Meteorological Society*, 92: 485–98. doi: 10.1175/BAMS-D-11-00094.1

Tebaldi, Claudia and Reto Knutti (2007). The use of the multi-model ensemble in probabilistic climate projections. *Philosophical Transactions. Series A, Mathematical, Physical, and Engineering Sciences*, 365(1857): 2053–75.

Thi, Phuong Cu and James E. Ball (2015). Estimating design flood magnitude for a Vietnamese catchment. In: *36th Hydrology and Water Resources Symposium: The Art and Science of Water, 1370*. Barton, ACT, Australia: Engineers Australia. Available at: https://search.informit.com.au/documentSummary;dn=824364044355354;res=IELENG

Tomassetti, Barbara, Marco Verdecchia, and Filippo Giorgi (2009). NN5: A neural network based approach for the downscaling of precipitation fields–model description and preliminary results. *Journal of Hydrology*, 367(1–2): 14–26.

Torma, C., F. Giorgi, and E. Coppola (2015). added value of regional climate modeling over areas characterized by complex terrain–precipitation over the Alps. *Journal of Geophysical Research, D: Atmospheres*, 120(9): 3957–72.

Trenberth, K. E. (2015). Has there been a hiatus? *Science*, 349: 691–2. Available at: http://dx.doi.org/10.1126/science.aac9225

UKCP18, United Kingdom Climate Projections (2018). Available at: www.metoffice.gov.uk/research/collaboration/ukcp/

Urwin, K. and A. Jordan (2008). Does public policy support or undermine climate change adaptation? Exploring policy interplay across different scales of governance. *Global Environmental Change: Human and Policy Dimensions*, 18(1): 180–91.

USGCRP (2000). National Assessment Synthesis Team, Climate Change Impacts on the United States: The Potential Consequences of Climate Variability and Change. Washington DC: US Global Change Research Program.

USGCRP (2016). US Global Change Research Program, Washington, DC.

USGCRP (2017). Climate Science Special Report: Fourth National Climate Assessment, Volume 1 [Wuebbles, D. J., D. W. Fahey, K. A. et al. (eds.)]. US Global Change Research Program: Washington, DC.

USGCRP (2018a). *Second State of the Carbon Cycle Report (SOCCR2): A Sustained Assessment Report* [N. Cavallaro, G. Shrestha, R. Birdsey, et al. (eds.)]. US Global Change Research Program, Washington, DC, 878 pp., doi.org/10.7930/SOCCR2.2018.

USGCRP (2018b). *Impacts, Risks, and Adaptation in the United States: Fourth National Climate Assessment, Volume II* [D. R. Reidmiller, C. W. Avery, D. R. Easterling, et al. (eds.)]. US Global Change Research Program, Washington, DC. doi: 10.7930/NCA4.2018.

Van Oldenborgh, G. J., Matthew Collins, Julie Arblaster, et al. (2013). Annex I: Atlas of global and regional climate projections. *Climate Change 2013: The Physical Science Basis Contribution of Working Group I to the Fifth Assessment Report of the Intergovernmental Panel on Climate Change* [Stocker, T.F., D. Qin, G.-K. Plattner, M. Tignor, S.K. Allen, J. Boschung, A. Nauels, Y. Xia, V. Bex and P.M. Midgley (eds.)]. Cambridge University Press, Cambridge, United Kingdom and New York, NY, USA.

Vano, Julie A., Jeffrey R. Arnold, Bart Nijssen et al. (2018). DOs and DON'Ts for using climate change information for water resource planning and management: Guidelines for study design. *Climate Services*, 12: 1–13.

VanRheenen, Nathan T., Andrew W. Wood, Richard N. Palmer, and Dennis P. Lettenmaier (2004). Potential implications of PCM climate change scenarios for Sacramento–San Joaquin River Basin hydrology and water resources. *Climatic Change*, 62(1): 257–81.

Vavrus, Stephen J. and Ruben J. Behnke (2014). A comparison of projected future precipitation in Wisconsin using global and downscaled climate model simulations: Implications for public health. *International Journal of Climatology*, 34(10): 3106–24.

Vigaud, N., M. Vrac, and Y. Caballero (2013). Probabilistic downscaling of GCM scenarios over Southern India. *International Journal of Climatology*. https://doi.org/10.1002/joc.3509.

von Trentini, Fabian, Martin Leduc, and Ralf Ludwig (2019). Assessing natural variability in RCM signals: Comparison of a multi model EURO-CORDEX ensemble with a 50-member single model large ensemble. *Climate Dynamics*, 53 (3–4): 1963–79. Available at: https://doi.org/10.1007/s00382–019-04755-8.

Vose, James M., James S. Clark, Charles H. Luce, and Toral Patel-Weynand, (eds.) (2016). *Effects of Drought on Forests and Rangelands in the United States: A Comprehensive Science Synthesis*. Gen. Tech. Rep. WO-93b. Washington, DC: US Department of Agriculture, Forest Service, Washington Office. 289 p.

Vose, R. S., D. R. Easterling, K. E. Kunkel, A. N. LeGrande, and M. F. Wehner (2017). Temperature changes in the United States. In D. J. Wuebbles, D. W. Fahey, K. A. Hibbard et al. (eds.), *Climate Science Special Report: Fourth National Climate Assessment, Volume I*. US Global Change Research Program, Washington, DC.: 185–206. doi: 10.7930/J0N29V45.

Vrac, M., M. Stein, and K. Hayhoe (2007). Statistical Downscaling of Precipitation through Nonhomogeneous Stochastic Weather Typing. *Climate Research*. Available at: https://doi.org/10.3354/cr00696.

Vu, Tue M., Ashok K. Mishra, Goutam Konapala, and Di Liu (2018). Evaluation of multiple stochastic rainfall generators in diverse climatic regions. *Stochastic Environmental Research and Risk Assessment: Research Journal*, 32(5): 1337–53.

Walsh, C. L., D., Roberts, R. J. Dawson, et al. (2013). Experiences of integrated assessment of climate impacts, adaptation and mitigation modelling in London and Durban. *Environment and Urbanization*, 25(2): 361–80. Available at: https://doi .org/10.1177/0956247813501121

Walton, D. B., F. Sun, A. Hall, and S. Capps (2015). A hybrid dynamical-statistical technique: Part I: Development and validation of the technique. *Journal of Climate*, 28: 4597–617

Wang, Jiali, Prasanna Balaprakash, and Rao Kotamarthi (2019a). Fast domain-aware neural network emulation of a planetary boundary layer parameterization in a numerical weather forecast model. *Geoscientific Model Development*, 12(10). Available at: https://doi.org/10.5194/gmd-2019-79

Wang, L., Chen, W., Zhou, W. (2014b). Assessment of future drought in Southwest China based on CMIP5 Multimodel projections. *Advances in Atmospheric Sciences*, 33: 1035–50.

Wang, B., G. Hong, C. Q. Cui, et al. (2019b). *Front. Eng. Manag.* 6: 52. https://doi.org/10 .1007/s42524–019-0002-y

Wang, J. and V. R. Kotamarthi (2014). Downscaling with a nested regional climate model in near-surface fields over the contiguous United States. *Journal of Geophysical Research, D: Atmospheres*, 119(14): 8778–97.

Wang, J. and V. R. Kotamarthi (2015). High-resolution dynamically downscaled projections of precipitation in the mid and late 21st century over North America. *Earth's Future*, 3(7): 268–88.

Wang, C., B. Lin, C. Chen, and S. Lo (2015b). Quantifying the effects of long-term climate change on tropical cyclone rainfall using a cloud-resolving model: Examples of two landfall typhoons in Taiwan. *Journal of Climate*, 28: 66–85. Available at: https://doi .org/10.1175/JCLI-D-14-00044.1

Wang, M., J. E. Overland, V. Kattsov, J. E. Walsh, X. Zhang, and T. Pavlova (2007). Intrinsic versus forced variation in coupled climate model simulations over the Arctic during the twentieth century. *Journal of Climate*, 20: 1093–107. Available at: http:// dx.doi.org/10.1175/JCLI4043.1

Wang, J. F. N., U. Swati, M. L. Stein, and V. R. Kotamarthi (2015a). Model performance in spatiotemporal patterns of precipitation: New methods for identifying value added by a regional climate model. *Journal of Geophysical Research: Atmospheres*. Available at: https://doi.org/10.1002/2014jd022434

Wang, C., L. Zhang, S.-K. Lee, L. Wu, and C. R. Mechoso (2014a). A global perspective on CMIP5 climate model biases. *Nature Climate Change*, 4: 201–5. Available at: http://dx.doi.org/10.1038/nclimate2118

Warren, F. J. and D. S. Lemmen., (eds.) (2014). *Canada in a Changing Climate: Sector Perspectives on Impacts and Adaptation*. Government of Canada, Ottawa, ON, 286p.

Warrick, R., W. Ye, Y. Li, M. Dooley, and P. Urich (2009). SimCLIM: A Software System for Modelling the Impacts of Climate Variability and Change. Hamilton, New Zealand: CLIMsystems Ltd.

Wasko, Conrad, Alexander Pui, Ashish Sharma, Rajeshwar Mehrotra, and Erwin Jeremiah (2015). Representing low-frequency variability in continuous rainfall simulations: A hierarchical random B Artlett L Ewis Continuous Rainfall Generation Model. *Water Resources Research*, 51(12): 9995–10007.

Wasko, Conrad, Ashish Sharma, and Fiona Johnson (2015). Does storm duration modulate the extreme precipitation-temperature scaling relationship? *Geophysical Research Letters*, 42(20): 8783–90.

Watson, A., J. Reece, B. E. Tirpak, et al. (2015). The Gulf Coast Vulnerability Assessment: *Mangrove, Tidal Emergent Marsh, Barrier Islands, and Oyster Reef*. 132 pp.

Available from: http://gulfcoastprairielcc.org/science/science-projects/gulf-coast-vul nerability-assessment/

Watson, Robert T., Marufu C. Zinyowera, and Richard H. Moss (1996). Climate change 1995 1996. Impacts, adaptations and mitigation of climate change: Scientific-technical analyses. In: Robert T. Watson, Marufu C. Zinyowera, Richard H. Moss and David J. Dokken (eds.), Climate change 1995–Impacts, adaptations and mitigation of climate change: Scientific-technical analyses. Cambridge: Cambridge University Press: 879 .

Weart, S. (2015). CLIMATE AND CLIMATE CHANGE | History of Scientific Work on Climate Change. *Encyclopedia of Atmospheric Sciences*. Available at: https://doi.org/ 10.1016/b978–0-12-382225-

Wehner, M. F., J. R. Arnold, T. Knutson, K. E. Kunkel, and A. N. LeGrande (2017). Droughts, floods, and wildfires. In: D. J. Wuebbles, D. W. Fahey, K. A. Hibbard, et al. (eds.), *Climate Science Special Report: Fourth National Climate Assessment, Volume I.* Washington, DC.: U.S. Global Change Research Program: 231–56.

Weigel, A. P., R. Knutti, M. A. Liniger, and C. Appenzeller (2010). Risks of model weighting in multimodel climate projections. *Journal of Climate*, 23: 4175–91. Available at: http://dx.doi.org/10.1175/2010jcli3594.1

Wigley, T. M. L., P. D. Jones, K. R. Briffa, and G. Smith (1990). Obtaining sub-grid-scale information from coarse-resolution general circulation model output. *Journal of Geophysical Research, D: Atmospheres*, 95(D2): 1943–53.

Wigley, T. M. L. and S. C. B. Raper (1990). Natural variability of the climate system and detection of the greenhouse effect. *Nature, 344*(6264): 324.

Wilby, Robert L., S. P. Charles, Eduardo Zorita, et al. (2004). Guidelines for Use of Climate Scenarios Developed from Statistical Downscaling Methods. *Supporting Material of the Intergovernmental Panel on Climate Change, Available from the DDC of IPCC TGCIA* 27. http://www.narccap.ucar.edu/doc/tgica-guidance-2004.pdf

Wilby, Robert L., Christian W. Dawson, and Elaine M. Barrow (2002). SDSM–a decision support tool for the assessment of regional climate change impacts. *Environmental Modelling & Software*, 17(2): 145–57.

Wilby, Robert L., Christian W. Dawson, Conor Murphy, P. O. Connor, and Ed Hawkins (2014). The Statistical DownScaling Model-Decision Centric (SDSM-DC): Concep-tual basis and applications. *Climate Research*, 61(3): 259–76.

Wilby, R. L. and T. M. L. Wigley (1997). Downscaling general circulation model output: A review of methods and limitations. *Progress in Physical Geography: Earth and Environment*, 21(4): 530–48.

Wilby, R. and T. M. L. Wigley (2000). Hydrologic responses to dynamically and statistic-ally downscaled climate model output. *Geophysical Research Letters*, 27: 1199–202.

Wilby, R. L., T. M. L Wigley, D. Conway et al. (1998). Statistical downscaling of general circulation model output: A comparison of methods, *Water Resource Research*, 34 (11): 2995–3008. Doi:10.1029/98WR02577

Wilks, D. S. and R. L. Wilby (1999). The weather generation game: A review of stochastic weather models. *Progress in Physical Geography: Earth and Environment,* 23 (3): 329–57.

Wilson, L. and W. Barnett (1983). Degree-days: An aid in crop and pest management. *California Agriculture*, 37(1): 4–7.

Woo, S., Singh, G. P., Oh, J. H. et al. (2019). Projection of seasonal summer precipitation over Indian sub-continent with a high-resolution AGCM based on the RCP scenarios. *Meteorology and Atmospheric Physics*, 131: 897. Available at: https://doi.org/10.1007/ s00703–018-0612-7

Wood, A. W., Lai R. Leung, Sridhar Venkataramana, and D. P. Lettenmaier (2004). Hydrological implications of dynamical and statistical approaches to downscaling climate model outputs. *Climatic Change*, 62: 1890216.

Wood, A. W., E. P. Maurer, A. Kumar, and D. P. Lettenmaier (2002). Long range experimental hydrologic forecasting for the Eastern U.S., *Journal of Geophysical Research*, 107(D20): 4429. Doi:10.1029/2001JD000659

Wuebbles, D., D. W. Fahey, and K. A. Hibbard (2017). How will climate change affect the United States in decades to come? *EOS*, 98. Available at: https://doi.org/10.1029/2017EO086015

Xu, Chong-Yu (1999). From GCMs to river flow: A review of downscaling methods and hydrologic modelling approaches. *Progress in Physical Geography: Earth and Environmen*, 23(2): 229–49.

Xu, Z. and Z.-L. Yang (2012). An improved dynamical downscaling method with gcm bias corrections and its validation with 30 years of climate simulations. *Journal of Climate*, 25(18): 6271–86.

Zeebe, Richard E., Andy Ridgwell, and James C. Zachos (2016). Anthropogenic carbon release rate unprecedented during the past 66 million years. *Nature Geoscience*, 9(4): 325.

Zhang, Yongfang, Dexin Guan, Changjie Jin, et al. (2011). Analysis of impacts of climate variability and human activity on streamflow for a river basin in Northeast China. *Journal of Hydrology*, 410(3): 239–47.

Zhang, G. J. and N. A. McFarlane (1995). Sensitivity of climate simulations to the parameterization of cumulus convection in the Canadian climate centre general circulation model, *Atmosphere-Ocean*, 33(3): 407–46. DOI: 10.1080/07055900.1995.9649539

Zobel, Z., J. Wang, D. J. Wuebbles, and V. R. Kotamarthi (2017). High-resolution dynamical downscaling ensemble projections of future extreme temperature distributions for the United States. *Earth's Future*, 5(12): 1234–51.

Zobel, Z., J. Wang, D. J. Wuebbles, and V. R. Kotamarthi (2018). Evaluations of high-resolution dynamically downscaled ensembles over the contiguous United States. *Climate Dynamics*, 50(3–4): 863–84. Available at: https://doi.org/10.1007/s00382-017-3645-6

Index

1-in-100-year, 125
1-in-100-year precipitation event, 125

a non-parametric, 88
A1FI, 131
abrupt, 126
absorption, 21
accelerated CPUs, 159
accuracy, 102, 160
accurate, 104
actionable information, 117
active flood management, 14
adaptation, 1, 4, 6–7, 10, 12, 14, 20, 41, 134, 137, 162, 164
adaptation actions, 137
adaptation and mitigation, 117
adaptation and resilience planning, 1
adaptation planning, 8, 104, 150
adaptation strategies, 122
adaptation strategy, 14
adaptive hierarchical, 160
add value, 100, 104, 109
added value, 80, 102–4, 106, 109, 111, 117, 119, 145, 162
Added value, 104–5
added value index (AVI), 106
advection, 69
advection and mixing, 26
aerosol particles, 21
aerosols, 21, 24, 29, 68
Africa, 141
agricultural systems, 48
agricultural yield, 140
agriculture, 1–2, 9–10, 141, 157
air pollution, 40
air quality, 9, 121, 132
Alaska, 14, 62
albedo, 26, 112, 126, 159
algal bloom, 2
Alps, 106

alternate planet, 33
Amazonia, 50
AMICAF, 48
amplify, 125
annual, 84
Antarctica, 159
anthropogenic, 27
appropriate scales, 150
appropriateness, 115
AR5, 3, 37
ArcGIS, 140
archived, 35
Arctic, 39, 126
Arctic sea ice, 36, 133
Argentina, 48
armoring of infrastructure, 14
Arrhenius, 20
ARRM, 91
ARRM dataset, 86
artificial intelligence, 83
Artificial intelligence, 159
artificial neural networks, 92
Artificial Neural networks, 98
Assessing uncertainty, 150
assessment, 38, 51, 54, 143
assessment of risk, 104
Assessment Report, 4
assessments, 3, 57, 157
Assessments, 40
associated uncertainty, 136
Asynchronous regression, 89, 91
Atlantic Multidecadal Oscillation, 125
atmosphere, 19, 29
atmospheric boundary layer, 27, 66
atmospheric chemistry, 24, 27
atmospheric circulation, 26, 67
Atmospheric Circulation, 26
atmospheric clouds, 21
atmospheric gases, 68
atmospheric processes, 128

atmospheric resolution, 30
Atmospheric Water Vapor, 73
Australasia, 124
Australia, 59
Australia(, 41
automated tool, 139
average biases, 38

Bangladesh, 54
Bartlett-Lewis, 93
Bay of Bengal, 55
Bayesian frameworks, 92
best model, 129
bias, 79, 164
bias correcting, 98
bias correction, 42, 74, 82–3, 87, 91, 96,
    102, 121
bias corrections, 91
Bias correction–spatial desegregation, 89
bias correction–spatial disaggregation, 91
bias-correct, 7, 82–3, 85, 146
bias-corrected, 6–7, 79, 83, 88, 147, 164
bias-correcting, 86
bias-correction, 43, 77, 82
Big Bend National Park, 141
binary format, 73
biogeochemical cycles, 27
biogeochemistry, 24–5
Biogeochemistry, 27
biological processes, 68
Biological processes, 21
biosphere, 19, 29, 68
boundary conditions, 79, 109
boundary layer, 26
boundary-value problem, 23
Brazil, 47
bridging the gap, 83
broad indicators, 152
broad-scale spatial trends, 109
Broward County, 16
brush fires, 60
building standards, 12
built environment, 122
Bureau of Meteorology, 59
business as usual, 133

calibration, 82
calibration period, 96
California, 41, 46, 115
Canada, 41, 46
Canada's Changing Climate Report, 46
canonical correlation analysis, 84
capital improvement plans, 11
carbon, 27, 159
carbon cycle, 24, 28–9, 125, 128
Carbon cycle, 27
carbon dioxide, 1, 20, 130
carbon emissions, 126, 133–4

carbon sink, 27
Cascade Range, 44
caution, 152
CDFt, 91
CDFt method, 91
cells, 21
Central Africa, 57–8
central part of the distribution, 98
certification programs, 142
CESM, 160
change points, 98
chaotic, 123
Charney Report, 4
Chicago, 46, 105
China, 51–2
circulation, 26
Cities, 160
city, 140
city planning, 141
civil engineering researchers, 119
climate change, 138
climate change impacts, 87, 119
climate change information, 150
climate change signals, 115
climate event, 125
climate events, 157
climate extremes, 42
climate factors, 122
climate impacts, 7, 40, 122
climate indicators, 83
climate information, 15, 150, 157
Climate information, 117
climate information websites, 153
climate interpretation, 142
climate interpreters, 142, 158
climate model, 82
climate modeling, 19, 25
climate models, 31, 142
Climate models, 36
climate projections, 23, 102–4, 107, 125, 131, 133,
    136, 141, 153, 163–4
climate realism, 145
climate scientists, 104, 119, 141
climate sensitivity, 75, 121, 125, 128, 137
Climate sensitivity, 125, 128
Climate Services, 153
climate state, 94
climate system, 126–7, 133
climate variability, 74, 82
climate variables, 102
climate-policy scenarios, 131
climate-ready, 117
climate-related risk, 13
climatic factors, 122
climatological periods, 96, 137
climatological statistics, 96
climatology, 82, 93
closed-form solutions, 24

cloud characteristics, 126
cloud droplets, 68
cloud formation, 26, 128
cloud properties, 133
Cloud Properties, 72
clouds, 29, 67, 128
Clouds, 25
clouds and aerosols, 128
clouds, 29
clustering methods, 92
CMIP, 101, 160, *See* Coupled Model Intercomparison
    Project
CMIP3, 36, 48, 51, 54, 131
CMIP5, 32–3, 36, 47, 51–2, 54, 56–7, 128, 131, 142
CMIP6, 28, 32, 37, 128, 131, 160
coarse-resolution GCM, 98
coastal adaptation, 14
coastal assets, 15
coastal climate, 104
coastal communities, 13
Coastal ecosystems, 14
Coastal Resilience, 16
coastal storm barriers, 14
coastal storms, 11, 15
coastal zones, 61
coastlines, 15, 66
Color coding, 152
Colorado, 46
commodity price shocks, 12
complex, 104, 122
complex coastlines, 106
complex systems, 19
complex terrain, 66
components, 127
computable, 103
computational cost, 30
computational costs, 84
computational frameworks, 29
Computational models, 19
computational resources, 95
computationally efficient, 24
computationally expensive, 65
computationally intensive, 160
computationally-intensive, 159
condensation, 26
conducting heat, 27
conservation improvements, 10
conservation of energy, 24
constructed analogue approach, 87
control simulation, 32
convection, 20–1, 26, 67, 106, 159
convection-permitting RCM, 106
convection-resolving models, 116
convection-resolving RCM, 103
convective, 70
convective cloud scheme, 74
convective mixing, 26
convective parameterization, 160

convective rainfall, 108
convective systems, 25
convolutional neural networks, 92
cooler than average, 124
cooling centers, 137
CORDEX, 80, 143, 153, 160
CORDEX-AFRICA, 57
COSMO-CLM, 57, 67
cost-prohibitive, 133
Coupled Model Intercomparison Project, 35
coupled processes, 162
coupling, 65
CPUs, 159
credibility, 115
credibility analysis, 106
credible, 103–5, 108
credible climate information, 144
credible information, 117
credible physical mechanism, 108
credibly, 144
critical infrastructure, 2
crop model, 105
crop yield, 40–1, 105
crop yields, 51, 121
cross-sectoral planning, 2
cryosphere, 29
CSIRO, 59
cumulative annual precipitation, 84
Cumulative density functions, 86
Cumulative distribution function transform, 89, 137
cumulus clouds, 70
current energy infrastructure, 132
customization, 104
cyclical patterns, 123
cyclones, 106

Da Nang, 8
daily average humidity, 91
daily maximum, 75
daily projections, 84
daily values, 85
daily weather pattern, 87
data format, 153
data quality, 155
data-simulation, 159
days below freezing, 105
day-to-day variability, 84
day-to-day variations, 88
decision making, 117, 141, 152
Decision Scaling, 15
decision support, 141
decision support tool, 87, 165
decision-makers, 43, 45, 104, 122, 132, 136
decision-making, 9, 105, 117, 126, 165
decision-making protocols, 15
decision-tree, 139
Deep and Shallow Convective Clouds, 26
deep convection, 106

deep convective, 68
deep neural networking, 159
deep neural networks learning, 92
deep ocean, 126
degree of appropriateness, 152
degree-days, 7
delta, 84, 90
Delta, 89, 99
delta approach, 111
delta correction, 99
delta method, 16
demand management, 13
demographic information, 140
demographics, 122–3, 130
desert, 59
deserts, 47
design and construction, 103
design floods, 120
detect, 98
digital coast mapper, 63
diminish, 125
disaggregate, 82, 88, 95
disaggregation, 95
disaggregation methods, 93
discretization, 24
discretize, 69
downscale, 7, 82
downscaled, 83, 88, 129, 164
downscaled climate projections, 121
downscaled data, 150
downscaled information, 150
downscaled simulations, 74
downscaling, 6, 43, 45, 77, 86, 101, 117, 135, 163
downscaling methods, 143
downscaling studies, 102
downscaling techniques, 42
downscaling, 117
drainage system, 1
drier than average, 124
drier-than-average, 124
drinking water, 12
drinking-water wells, 14
drizzling, 26
droplet growth, 68
drought, 117, 140
drought risk, 134
droughts, 1, 11–12, 51, 58
dry days, 47
duration and intensity, 93–4
dynamic downscaling, 16, 62, 81
Dynamic downscaling, 44
dynamic high-resolution global model simulation, 99
dynamic vegetation, 68
dynamical, 144
dynamical downscaling, 66, 103, 105, 120, 146
Dynamical downscaling, 7, 70
dynamical model, 111
dynamical weather, 82

dynamically, 164
dynamically downscaled, 16, 73, 75, 80
dynamics of the system, 122

Earth System Models, 24, 29
Earth System Sensitivity, 126
East Asia, 51–2
East Asian monsoon, 52
easterlies, 27
Eastern Australia, 61
eastern Sahel, 58
ecological regions, 45
ecologists, 140
economics, 2
ecosystem, 105
ecosystem health, 132
ecosystem services, 14, 40
ecosystems, 1–2, 15, 83, 122, 157
Ecosystems, 40
ECS, 126
EDQM, 91
EDSM, 104–5
efficiency, 146
El Nino, 32
El Niño, 124, 135
El Niño-Southern Oscillation, 124
electric cars, 133
eliminate, 137
eliminating carbon emissions, 133
embed, 65
Emission Scenarios, 142
emissions, 1, 126, 158
emperature, 110
empirical equations, 24
empirical quantile mapping, 85–6
Empirical quantile mapping, 85
empirical statistical downscaling, 82–3
Empirical statistical downscaling, 7
empirical-statistical, 82, 103
empirical-statistical downscaling, 82
Empirical-statistical downscaling models, 44
emulation, 159
end user, 104
endangered, 140
energy, 1, 69
energy balance models, 28
Energy Balance Models, 20
energy budget, 21
energy flow, 159
energy infrastructure, 122, 133
energy production, 132
energy supply, 2
energy transfer, 69
energy use, 130
Energy utilities, 10, 13
engaging users, 154
engineering design, 119
Engineering design, 120

engineers, 136
ensemble, 75, 129
ENSEMBLE, 163
ensembles, 36–7, 143
ENSO, 33, 84, 94
entire distribution, 84
entire statistical distribution, 91
entrainment, 68
epidemiological model, 105
Equidistant quantile mapping, 90
equilibrium climate sensitivity, 126
error-checking, 147
ESDM, 45, 48, 82–7, 92, 96, 98, 101–2, 109, 111–13, 115, 145–7, 149, 158, 162
ESDM stationarity, 111
ESS, 126
EURO-CORDEX, 50, 153
Europe, 50, 107, 153
European Economy, 42
European Union, 41
evaporation, 24, 26–7
every location, 87
evolving, 127
exascale, 157–8, 164
expansive coastal wetlands, 14
external expertise, 153
external forcing, 29
extratropical cyclones, 15
extreme, 125
extreme case of adaptation, 143
extreme conditions, 82
extreme downpours, 12
extreme drought, 31, 49
extreme events, 46, 106, 117
Extreme events, 52
extreme heat, 12, 46, 105, 117, 134–5
Extreme heat, 11, 55
extreme precipitation, 47, 95, 140
Extreme precipitation, 55
extreme rainfall, 94
extreme tails, 98
extreme temperature, 140
extreme value distributions, 94
extreme weather, 11, 157
extreme weather events, 2
extremes, 85, 101, 111, 129, 152

failure to act, 136
Federal Highway Administration, 11
Fifth IPCC Assessment Report, 131
fine spatial scale, 87
finer resolutions, 31
finer scale phenomena, 106
finer spatial scales, 125
First National Climate Assessment, 5
first ranks, 85
fisheries, 41
flexibility, 146

flood, 122
flood management, 10
flood maps, 12
flood risk, 12, 92, 134, 140
flood risk management, 92
flooding, 12
floods, 2, 12
fluid motion, 24
food insecurity, 12
food security, 121
for impacts and adaptation, 143
forecast, 83
forest fires, 47
forest resource management, 10
forestry industry, 119
form of the distribution, 88
Fort Lauderdale, 130
fossil fuel, 126
fossil fuels, 34, 68, 132, 140
fossil-fueled development, 131
fossil-fuel-intensive, 130
fossil-intensive, 131
Fourth National Climate Assessment, 42, *National Climate Assessment*
France, 50
freeze-thaw, 119
frequency, 11
frequency and intensity, 129
frequency, 117
fuel availability, 61
future emissions, 130, 135
future projection, 100
future projections, 35, 108, 149
future scenario, 95
future scenarios, 75, 134, 141
Future scenarios, 34

gamma, 88
gamma distributions, 94
gases, 21
gas-powered, 133
gaussian, 88, 92
GCM, 6, 22, 24, 26, 28, 33, 64, 74, 83–5, 87, 95, 98, 101–6, 115, 126–7, 129, 137, 143–4, 146, 150, 158–9, 164
GCM projections, 100
GCM weighting schemes, 129
GCM, 103
GDP, 42
General Circulation Models, 21
general recommendations, 152
general suitability, 152
generating weather statistics, 100
geographic features, 152
Geographic Information Systems, 164
geographic location, 135
Geographical Information System, 43
geo-political, 45

geospatial, 73
geospatial data, 165
Germany, 50
gigabytes, 75
glacial maximum, 62
Glacial Maximum, 32
glacier melt, 56
glacier melt water, 56
global average changes, 43
global climate model projections, 11
global climate models, 6, 19, 21, 31, 33, 82
global commerce, 12
global coverage, 153
global mean sea-level, 62
global mean temperature, 134
global model, 153
global targets, 134
global temperature, 33
goodness-of-fit, 96, 109
goods and services, 12
governing equations, 69
GPU, 159
gravitational field, 62
gravity, 69
gravity flow, 27
Great Lakes, 41
Great Plains, 109
green, 152
greenhouse gas, 162
greenhouse gases, 1, 50, 143
Greenland, 159
Greenland and Antarctic ice sheets, 126
Greenland or Antarctic ice sheets, 133
grid cell, 69, 87
grid cells, 69
grid size, 69
gridded dataset, 140
gridded observations, 7
grid-point free-atmosphere, 83
ground water table, 27
groundwater table, 17
growing degree-days, 92
growing-season precipitation, 84
guarantees, 153
guidance, 145, 155
Gulf Coast, 41, 61
Gulf of Mexico, 15
Gulf Stream, 21
Gumbel, 94

HadRM3P, 67
health, 1
health sector, 13
healthcare infrastructure, 10
heat islands, 12
heat maps, 140
heat stress, 11
heat stroke, 11

heat waves, 105
heat-driven mixing, 68
heat-trapping gas emissions, 34
heat-trapping gases, 126, 130, 133, 137
heatwaves, 11, 137
heavy downpours, 11
heavy precipitation, 116, 135, 138
heavy rain events, 117, 122
hierarchical Bayesian model, 94
high albedo, 28
high resolution, 64
high topographical information, 93
high variability, 94
higher latitudes, 111
higher model resolution, 160
higher resolution, 30, 65, 158
higher resolution regional model, 105
higher scenario, 134, 137
higher sensitivity, 128
higher-resolution, 92
higher-resolution information, 7
higher-resolution simulations, 116
higher-resolution spatial, 102
higher-resolution temporal, 102
highest RCP, 143
high-resolution, 42, 102, 164
high-resolution climate projections, 16, 83, 122, 129
high-resolution GCM, 98
high-resolution grid, 105
high-resolution modeling, 123, 128
high-resolution observations, 162
high-resolution projections, 82, 84, 129, 135
high-resolution spatial, 82
Himalayas, 52, 54, 56, 112
Hindu Kush, 56, 112
historical data, 98
historical diagnostics, 163
historical instrumental data, 82
historical observations, 82, 85, 109
historical period, 96
historical records, 157
historical sub-daily precipitation, 93
Holocene, 32
horizontal edges, 65
horizontal grid spacing, 30
horizontal resolution, 22, 31
horizontal spatial scales, 31
hourly, 103
hourly values, 103
human and natural system, 1
human choices, 20, 130
human civilizations, 157
human factors, 162
human health, 141
Human health, 40
human-centric, 45
humidity, 105, 110, 140
hybrid, 80

hybrid approach, 86, 164
hydraulic engineer, 103
hydraulic engineering, 92
hydraulic modeling, 94
hydrocarbons, 28
hydrological assessments, 91
Hydrological Cycle, 26
hydrological modeling, 93
hydrology, 66
hydrostatic assumption, 69

ice sheet, 24, 128
ice sheet dynamics, 29
ice sheet melt, 135
ice sheets, 28, 62
impact, 43
impact analysis, 103
impact assessments, 1, 70, 84, 102, 150
impacts, 4, 7, 157
impacts and adaptation, 143
impacts and adaptation planning, 142
importance, 133
independent models, 130
independent observations, 97
India, 54, 124
Indian Ocean, 59
indicators, 7
inequality, 131
inertia in the climate system response, 133
inevitable, 132
infiltration, 27
inform standards, 9
information broker, 142
information fit, 152
infrared radiation, 27–8
infrastructure, 2, 31, 61, 103, 117, 122, 125, 137, 140
Infrastructure, 40
Infrastructure & Climate Network, 119
infrastructure design, 9, 92, 119
infrastructure planning, 13
initial conditions, 23, 123
initial state, 126
initial state problem, 23
initial warming, 125
inland bodies of water, 106
input fields, 74
insensitive, 137
integrating analysis, 165
intensity, 11, 117
intensity and duration, 93
intensity of heat waves, 122
Intensity-Duration-Frequency, 94
interaction, 152
interactive biosphere, 162
interacts, 152
interface, 141
internal, 123
internal variability, 33, 125, 135

international supply chains, 2
interpretation of data, 155
intertropical convergence zone, 57
invasive species, 140
IPCC, 3–4, 141, 145, 153
IPCC AR 4, 54, 58
IPCC AR 5, 52, 57
IPCC AR 5, 52
IPCC AR 6, 129
IPCC assessment report, 36
IPCC Assessment Reports, 131
IPCC Second Assessment Report, 45
IPCC-AR 2, 6
irreversible changes, 126
irrigation, 10

jackknifed cross-validation, 96
Japan, 51
jet stream, 21, 30
jet streams, 27

Kernel, 90
kernel density distribution mapping, 91
kernel density estimation, 88
knowledge, 152

La Niña, 124
lake, 140
lake effects, 163
lake-effect snows, 163
land, 19, 24
land cover, 66
land cover and land use, 123
Land Surface Representation, 27
land surfaces, 29
land use, 9, 44, 66, 126
land-based ice, 61
landcover, 44
large biases, 111
large lakes, 104
large-scale circulation, 30
large-scale dynamics, 129
large-scale predictor, 84, 149
large-scale predictors, 82, 86, 98
large-scale upper air variables, 91
large-scale weather systems, 24
latent heat flux, 70
layers of soil, 27
lifecycle, 21, 26
lifecycle of carbon, 27
limitations, 11, 152
limited area, 65
limited by observations, 147
limited-area model, 65
linear, 88
linear combination, 81
linear multiple regression, 91
livestock, 40

LOCA, 87, 115
local, 40, 82
local change, 109
local climate, 86
local impact variable, 91
local scale, 43, 82
local surface observations, 83
local to regional scale, 101
localization, 104
Localized Constructed Analogues, 42
log-Pearson II, 94
longer timescales, 135
long-term droughts, 58
long-term storage, 27
long-term temperature, 135
long-term trend, 88
long-term trends, 140, 149
long-wavelength, 68
look-up table, 114
Low Elevation Coastal Zone, 13
lower scenario, 134
low-level clouds, 26
low-pressure, 113

MACA, 87
machine learning, 159, 162
magnitude of extreme events, 11
maps, 110
marine ecosystems, 25
marine environments, 26
mass, 24
maximum streamflow, 112
maximum temperature, 112
mean, 82
mean absolute error, 111
mean summer temperature, 84
mean values, 103
measurement, 165
mechanical, 68
Mediterranean, 50
MELODIST model, 95
mesonets, 163
mesoscale convective systems, 116
Mesoscale Modeling version 4, 67
mesosphere, 27
meta data, 154
metafile, 73
methane, 1, 27, 159
Method of Fragments, 94
metrics, 76, 106, 129
Mexico, 48
microclimates, 44
MicroMet package, 95
microphysical, 68
microphysics, 26
microscale, 24
microscale process, 26
mid-century, 135

middle of the road, 131
mid-range scenario, 136
migration, 51, 165
military installations, 141
millennia, 24
million core hours, 75
mitigation, 20
mitigation pathways, 4
mixing process, 66
model domain, 74
model independence, 38
Model Output Statistics, 83
model sensitivities, 29
model structure, 128
model time step, 22
model weights, 38
modeled accurately, 127
model-quality metrics, 38
moisture advection, 113
momentum, 24
monthly, 84
Monthly bias correction, 89
monthly mean local station data, 83
months to millennia, 126
more likely, 133
MOS, 84
multi-decadal, 23
multi-decadal analyses, 164
multi-decadal climate statistics, 33
multi-model ensemble, 137
multiple decades, 124
multiple future pathways, 140
multiple orthogonal values, 94
multiple realizations, 143
multi-variate, 162
Multivariate Adapted Constructed Analogs, 110
multi-variate ESDM, 87

NA-CORDEX, 80
NARCCAP, 90
national assessments, 41
National Climate Assessment, 5, 46
National Climate Assessment documents, 9
national security, 41
natural cycles, 135
natural emissions, 27
natural modes, 50, 124
Natural resources, 41
natural system, 104
natural variability, 29, 32, 96, 121, 123, 125, 135, 137,
    143, 158
Natural variability, 143
naval bases, 61
NCA4, 46, 129
ncentives, 133
nearest k-neighbor algorithm, 94
near-surface, 70
negative phase, 124

nested bias correction, 89, 91
net carbon emissions, 131
net negative, 130
net negative emissions, 131
netCDF, 140
NetCDF, 73, 153
Network Common Data Form, 73
neural networking algorithms, 92
New Hampshire, 119
New York City, 46
nitrogen, 27, 29
nitrogen cycle, 27
nitrous oxide, 27
noise, 96
non-climatic drivers, 123
non-linear processes, 123
non-native species, 9
non-parametric, 88
non-parametric distribution mapping method, 112
non-parametric method, 94
North Africa, 58
North America, 39, 45
North American Regional Scale Climate Change Assessment Project, 80
North Atlantic Oscillation, 125
North China Plain, 52
Northern Australia, 59
nudging, 74
Nudging, 74
Numerical solvers, 69
nutrients, 21

observational, 103
observational data, 125, 141
observational datasets, 158
observational record, 33
observations, 28, 82, 97–8, 103, 126, 128
observed climatology, 149
observed daily weather, 87
observed distribution, 91
observed patterns, 87
ocean circulation, 126
ocean currents, 28, *See* ocean
ocean surface, 68
oceans, 19
of carbon dioxide, 131
Okinawa, 52
one-dimensional, 20
outside the range projected, 126
overall uncertainty, 135
overestimate, 126
overfitting, 96
over-fitting, 86
over-investment, 136
over-preparation, 136–7
overturning circulation, 28
ozone, 28

Pacific Decadal Oscillation, 32, 125
Paleocene-Eocene Thermal Maximum, 126
paleoclimate, 19, 126–7
parameterization, 24
parameterizations, 128
parameterize, 158
parameterized, 21, 26–7, 70, 128
parametric, 75, 158, 164
parametric distributions, 88
parametric equation, 91
parametric quantile mapping, 85, 109
Parametric quantile mapping, 85
parametric uncertainties, 129
parametric uncertainty, 35, 70, 121, 125, 128
Parametric uncertainty, 128
parametrization, 128
Paris Agreement, 133–4, 137
partial differential equations, 24
past climates, 38
patterns, 82
pattern-recognition, 162
Pearson correlation coefficient, 77
peer approach, 153
peer-reviewed literature, 129
perfect model, 98, 111–12, 163
perfect prognosis, 98
perfectly accurate, 147
permitted, 119
perturbations, 29
Peru, 48
petascale, 158
phosphorous, 27
photosynthesis, 21
physical phenomena, 24
physical plausibility, 145
physical processes, 66, 105, 109, 128, 149
physically-based, 64
physics option, 74
physics packages, 75
physics parameterization, 29
piecewise linear regressions, 91
planetary scale waves, 27
planners, 136
planning standards, 9
policies, 133
policy, 157
policy changes, 133
policy choices, 130
policy makers, 119
poorly understood, 104
populatio, 15
population, 122
population growth, 165
portal, 154
portal design, 154
ports, 61
positive phase, 124

potential added value index (POV), 106
power plants, 133
practitioners, 141
precipitation, 10, 26–7, 30, 40, 46, 70, 102–3, 105–6, 108, 110, 113, 115, 128
Precipitation, 27
precipitation autolag, 91
precipitation bias, 79
precipitation extremes, 106, 111, 120
precision, 102
predictand, 87
predictands, 91
predictor, 84, 91
predictors, 83
predict-then-act, 15
pre-industrial levels, 133
pre-industrial revolution, 32
pre-industrial times, 126
prescribed forcing, 32
pressure systems, 67
primitive equations, 69
principle components, 84
probabilistic approach, 86
probabilistic frameworks, 163
probability density, 88
probability density functions, 77
Probability distributions, 110
process scale, 113
process-based evaluation, 163
process-based limitations, 163
process-based models, 128
processes, 127
process-level, 112, 115
process-level evaluations, 112
process-oriented, 106
Program for Climate Model Diagnosis and Intercomparison, 37
projected changes, 134
projected impacts, 105
proxy records, 32
public health, 2, 140
public safety, 2
purpose-built, 159

q-q relationships, 91
qualitatively, 41
quality of the data, 153
quantifying, 121, 135
quantifying impacts, 85
quantile mapping, 111
quantile of the distribution, 85
quantile of the GCM distribution, 85
quantile regression, 111
quantile-quantile plot, 86
quantile-quantile relationship, 91
quantiles of the distribution, 85
quantitative, 129
quantitative climate projections, 122–3

quantitative projections, 123, 138
quantitatively, 42
Quebec, 110

radiation, 24
radiative transfer, 29, 69–70
Radiative-convective models, 20–1
rain, 70
rainfall deficit, 50
rainfall intensity, 94
rainforest, 47
random cascade method, 95
random cascade model, 94
Random cascade models, 94
random number generator, 93
random variability, 96
randomly generate, 93
rapidly varying topography, 98
RCM, 46, 52, 69, 74, 76, 83, 87, 95–6, 101–2, 104–6, 112, 115, 145–6, 150, 158, 164
RCMs, 69, 79
RCP, 37, 131, 134, 142, 150
RCP 4.5, 16
RCP 8.5, 16
RCP2.6, 50, 131
RCP2.6/SSP126, 133
RCP4.5, 46, 48, 52, 55, 57–8, 75, 80, 131
RCP4.5/SSP245, 134
RCP6.0, 131
RCP7.0/SSP370, 158
RCP8.5, 46, 48, 50, 52, 54–5, 57–8, 60, 75, 80, 131, 133
RCP8.5/SSP585, 134, 158
reanalysis, 82
reanalysis data, 99
reanalysis dataset, 99
reflected, 28
RegCM2, 67
regional and local planning, 9
regional average values, 43
regional climate, 40
regional climate change, 40
regional climate model, 95, 143
regional climate model output, 95
regional climate modeling, 105–6
regional climate models, 44, 82, 161
Regional climate models, 64
regional climatic changes, 4
regional downscaling, 157
regional dynamics, 149
Regional Earth System Models, 161
regional economy, 42
regional impacts, 10
regional modeling, 164
regional phenomena, 113
regional risk assessment, 130
regional rivalries, 131
regional scale, 40, 102, 129

regional scales, 43
regional topography, 15
regional-scale impacts, 82
regional-scale ocean model, 62
regression + stochastic method, 87
regression equations, 83
regression model, 96
regression techniques, 84
re-gridding, 102
regulatory standards, 9
relative contribution, 135
relative humidity, 93
Relative sea level, 15
reliability, 137
reliable, 129
REMO-RCM, 67
renewable and clean energy sources, 10
Representative Concentration Pathways,
    131
reservoir, 140
resilience, 7, 11, 134
resilience planning, 6
resilience strategies, 137
resilience, 104
resilient infrastructure, 130
RESM, 161–2
resolution, 29
resolution of observations, 101
resolved, 104
resource constraints, 150
retreat, 14
return flows, 122
return period, 7
rising temperatures, 10
risk, 163
risk and adaptation, 141
risk envelope, 104
risks and impacts of climate change, 133
road and bridge design, 119
robust, 103–4
Robust Decision Making, 15
robustness, 38, 153
roof-top solar, 13
root mean square error, 110
root-mean-square error, 77
RSM, 147

sability, 152
Sahara, 56
salinity, 28
Sampling, 142
satellite data, 162
satellite datasets, 101
satellite observations, 82
scattering, 21
scenario, 133
scenario selection, 141
scenario uncertainty, 75, 133–5, 138

Scenario uncertainty, 130, 135
scenario-based, 16
scenario-based, 8
scenario-based planning, 8
scenarios, 75, 84, 132
Scenarios Networks for Alaska-Arctic and Planning,
    90
science of climate change, 20
scientific quality, 154
scientific tools, 157
scientific uncertainty, 127–8, 135
Scientific uncertainty, 125, 128, 135
SDSM, 91, 115
sea ice, 24, 28, 159
sea level, 61–2, 137, 157
sea level rise, 2, 10–11, 13–16, 40, 61, 119,
    134–5, 137
seasonal, 84
seasonal average precipitation, 129
seasonal climatology, 88
seasonal mean values, 152
seasonal temperatures, 105
Second National Climate Assessment, 5
Second State of the Carbon Cycle Report, 5
second-order Markov chain, 94
sector, 139
sector-based analysis, 122
selection, 155
semi-arid, 59
senescence, 27
sensitive, 122
shallow cumulus clouds, 68
Shared Socioeconomic Pathways, 37, 131
shifting seasonality, 117
short term, 135
short term drought, 58
shorter time scales, 125
shorter-term variations, 33
shortest atmospheric wave, 27
short-term variations, 33
shortwave, 28
shortwave energy, 28
short-wavelength, 68
signal decomposition, 88
signal processing, 88, 162
simulated patterns, 108
simulated sub-daily precipitation, 93
Sixth IPCC Assessment Report, 131
size infrastructure, 120
skill, 39, 106, 137
skill metric, 39
smaller spatial scales, 135
smaller-scale processes, 128
small-scale variations, 83
snow, 70
Snow, 71
snow albedo feedback, 115
snow melt, 12

snowmelt, 112, 120
snowmelt runoff, 120
snowpacks, 120
social justice issues, 40
social scientists, 119
socioeconomic, 4, 15, 105
socio-economic, 122
socio-economic, 162
socio-economic factors, 123
soil, 24
soil moisture, 27, 67
Soil moisture, 46
Soil Properties, 72
solar energy, 28
solar heating, 27
solar irradiance, 32
solar radiation, 26, 33, 140
South Africa, 57–8
South America, 47–9, 141
South Asia, 54
South Florida, 130
southeast Asia, 141
Southeast Asia, 124
Southeast Florida, 16
southeast winter monsoon, 54
Southern Africa, 58
Southern Australia, 59
Southern Ind, 56
Southern India, 55
Southwest Australia, 59
southwest summer monsoon, 54
spatial and temporal disaggregation, 82
spatial consistency, 87
spatial disaggregation, 83, 98
Spatial disaggregation, 91
spatial distribution, 106
spatial downscaling, 82, 87
spatial interpolation, 102
spatial resolution, 29, 66, 101
Spatial Resolution, 140
spatial scale of the observations, 86
spatial scales, 29, 128
spatially and temporally interpolate, 95
spatially disaggregate, 83
Spatiotemporal, 77
Special Report on Emission Scenarios, 131
specific station, 100
spring flooding, 120
spring thaw periods, 119
SRES, 36, 131, 134
SRES A1FI, 131
SRES B1, 131
SSP126, 131
SSP245, 131
SSP370, 131
SSP585, 131, 133
SSP-RCP, 134
SSPs, 37, 131–2

stakeholder, 102, 159
stakeholders, 42, 45, 115, 165
stakeholders, 141
stand-alone, 65
standard deviation, 77, 84, 91
state-of-the art models, 103
stationarity, 98, 111, 149, 162
stationary, 98
statistical, 144
Statistical Analysis of Residual Trends, 88
statistical bias-correcting, 62
statistical corrections, 82
statistical downscaling method, 82
statistical downscaling methods, 83
Statistical Downscaling Methods, 89
Statistical downscaling model, 90
Statistical DownScaling Model, 87, 91
statistical properties of the distribution, 86
statistical relationships, 80
statistical resampling techniques, 94
statistical sample, 37
statistical techniques, 16
statistical-dynamical approaches, 92
statistically, 164
stepped forward, 22
stochastic disaggregation, 93
stochastic disaggregation method, 93
stochastic methods, 86
stochastic model, 93
stochastic probability models, 83
Stochastic Weather Generators, 86
stochastically disaggregated, 103
storage of heat, 27
storage requirements, 75
storm sewer system, 138
storm surge, 15
storm surges, 12
storms, 106
stormwater, 2, 13, 16, 130, 165
Stratiform Clouds, 26
stratosphere, 27
stratospheric ozone, 29
streamflow, 7, 110
structural, 35, 75, 121, 129, 158, 164
structural uncertainties, 129
structural uncertainty, 70, 125
Structural uncertainty, 127
sub-daily, 103
sub-daily changes, 85
Sub-daily humidity, 95
sub-daily precipitation, 93, 103
sub-daily temporal information, 92
sub-grid, 25, 66
sub-grid scale, 91
Sub-grid scale processes, 128
sub-grid-scale processes, 128
sub-micron, 26
sub-model, 128

sub-models, 128
sub-regional, 150
subsidence, 61
substructure flooding, 2
subtropical, 51, 54
successful climate information, 153
suitability, 152
summary for policymakers, 3
summer monsoon, 55
summer precipitation, 107
sun, 28
sunspot activity, 28
supply chain, 12
support vector machines, 92
surface hydrology, 159
surface layer of the soil, 27
surface observations, 82
surface runoff, 12, 68
Surface runoff, 27
surface temperatures, 1
surface water storage, 14
surface weather conditions, 83
sustainability, 11, 131
SWG, 94

tails of the distribution, 87, 95
Taylor diagram, 77
Technical Capacity, 140
technological and economic inertia, 133
technological development, 130
temperature, 40, 70, 102, 105, 108, 112
temperature gradient, 28
Temperature Profile, 71
temperature records, 33
temperature trends, 38
temporal disaggregation, 87, 92–3, 95
temporal projections, 82
temporal resolution, 139
temporary slowing, 33
terabytes, 75
thawing permafrost, 126, 159
theoretical distribution, 125
theoretical extreme value distributions, 94
thermal expansion, 61
Third National Climate Assessment, 5
threat multiplier, 121
three-dimensional, 162
threshold exceedances, 7
time averages, 152
time scale, 135
time scales, 126
timescales, 128
tipping point, 14
topographical features, 104
topography, 27, 62, 105
tornado alley, 123
tornado outbreaks, 103
tornado risks, 104

tornadoes, 123
total forcing, 35, 37
tourism, 41
trace atmospheric gases, 27
training data, 115
training period, 7, 96
transfer function, 91
transferability, 96–7, 109
transient, 32
translating climate modeling, 9
translation, 104
translation tools, 153
transport, 69
transport of heat, 126
transportation, 40
transportation disruptions, 12
transportation infrastructure, 14, 119, 141
transportation sector, 117
treatment costs, 13
tree cover, 49
trends, 46
tropical, 54, 59
tropical climate, 57
tropical rainforests, 56
tropics, 26
troposphere, 27, 70
tropospheric temperature, 33
trucks, 119
trustworthy, 102
turbulence, 128
Type I error, 136
Type II error, 136
typhoons, 52

U.S. National Climate Assessment, 84, 87
U.S. National Climate Assessments, 131
UK, 50
unavut, 141
uncertainties, 92, 95, 108, 152, 157–8
uncertainty, 8, 15, 95, 103, 117, 122–3, 125, 132, 135, 137–8, 142
Uncertainty, 125
uncertainty assessment, 152
uncertainty, 160
underestimate, 126
underestimate temperature change, 126
underestimate the uncertainty, 136
under-preparation, 136–7
United Kingdom, 41
United States, 41, 46, 124
unknown, 103
unnecessary expenditures, 136
unpredictable, 123
unprepared, 136
unresolved scales, 128
uplift, 61
upper air patterns, 83
upper limit, 133

urban electricity demand, 92
urban environments, 29
urban heat island, 105, 140
urban population, 137
urban systems, 165
usability, 153
useable, 152
USGCRP, 4, 141

VALUE, 99, 112, 114
vapor pressure deficit, 24
variability, 74, 77, 140
variable-resolution, 7, 65, 81
Variable-resolution, 65
variance, 91
variance decomposition, 106
vegetation, 21, 24, 68
vehicles, 133
vertical resolution, 31
vertical structure, 29
very high resolutions, 164
very high-resolution, 152
violated, 98
visualizations, 165
volcanic eruptions, 32
vulnerability, 4, 6–7, 15, 117, 134
vulnerable, 105

warming climate, 9
warming patterns, 80
wastewater systems, 2
water and energy, 121
water demand, 132
water management, 105
water management model, 105
water quality, 9, 122
water resources managers, 145
water supply and demand, 140
water supply diversification, 10
water vapor, 26
watershed, 10
watershed hydrology model, 105
wave motions, 27
wavelength, 69

weather, 26
weather and climate extremes, 121
weather forecast models, 67, 82
weather forecasts, 23
weather generator, 102
weather generators, 99
weather model output, 83
weather models, 83
weather prediction, 160
weather prediction models, 82
weather processes, 160
weather records, 87
Weather Research Forecast, 67
weather station, 82, 87, 141
weather stations, 105, 115, 164
weather-prediction models, 64
Web Portals, 153
web sites, 153
weighted, 38
weighting schemes, 129
well-mixed slab, 21
well-tested GCM, 129
West Africa, 58
westerlies, 27
western Sahel, 58
wet bulb, 52, 55
wetter than average, 124
wetter-than-aver, 124
wildfire, 2, 134
wildfires, 9, 12
wind, 67
Wind, 72
wind speed, 24, 110
winter monsoon, 52
winter temperatures, 119
winter tourism, 141
World Climate Research Programme,
    35
WorldClim database, 84
WRF, 69
WRF dynamical simulations, 115
WRF regional model, 115

Yangtze River valley, 115